Structure and Retention in Chromatography

Chromatography: Principles and Practice

This book series will be a multi-volume, comprehensive treatise on chromatography to meet the requirements of chromatographers and nonspecialists who need to understand and use chromatography.

Edited by B. Ravindranath, Bharavi Laboratories Ltd., Bangalore, India

---

Volume 1
Structure and Retention in Chromatography: A Chemometric Approach
Roman Kaliszan

Other Volumes in Preparation

Supercritical Fluid Chromatography and Extraction
edited by *M. Caude and D. Thiebaut*

Planar Chromatography (TLC) – Handbook of Thin Layer Chromatography
edited by *E. Soczewinski*

Chromatographic Separation of Carbohydrates
edited by *S. Churms and A. Stephen*

Chromatographic Separation of Lipids
edited by *A. Kuksis*

This book is part of a series. The publisher will accept continuation orders which may be cancelled at any time and which provide for automatic billing and shipping of each title in the series upon publication. Please write for details.

Structure and Retention in Chromatography

A Chemometric Approach

Roman Kaliszan

Medical University of Gdańsk
and University of Gdańsk
Gdańsk, Poland

harwood academic publishers

Australia • Canada • China • France • Germany • India • Japan
Luxembourg • Malaysia • The Netherlands • Russia • Singapore
Switzerland • Thailand • United Kingdom

Printed in Singapore.

Amsteldijk 166
1st Floor
1079 LH Amsterdam
The Netherlands

British Cataloguing in Publication Data

Kaliszan, Roman
Structure and retention in chromatography : a chemometric approach. – (Chromatography : principles and practice ; v. 1)
1. Chromatographic analysis
I. Title
543′.089

ISBN 90-5702-028-9

Contents

Preface to the Series

Chromatography is well known to all practising chemists and biochemists as a powerful laboratory technique for separation and qualitative and quantitative analysis. However, it is not generally recognised what the underlying physicochemical principles are, how versatile the technique is, and how varied its applications. Over the past half century, chromatography has become an indispensable tool in several fields of science and technology. Indeed, it has changed the course of development of many of these fields, including biology, medicine and chemical technology. The same is true of progress in natural products chemistry, where developments in chromatography have directly contributed to exponential growth of the number of new chemical entities isolated and identified. The role of chromatography in analytical quality control and ecological sciences is well recognised. Biotechnology is another field which acutely depends on chromatography for its success; this is because biotechnological processes often yield complex mixtures of very closely related products and, the availability of efficient separation processes is a limiting factor in the successful use of the process.

The versatility of chromatography as a separation and analytical technique is immense. All types of compounds, from the small hydrogen and helium molecules to large and complex protein molecules, can be separated by chromatography. It is not exaggerated to state that there are no two compounds, however similar in structure (even optical isomers), which cannot be separated by one chromatographic technique or another. The tools available range from a simple laboratory microslide to the most sophisticated automated instrument and the methods, ranging from liquid–liquid partition to molecular recognition, are unlimited. Presumably no other technique also offers a range in scale of operation from attogram to gigagram!

Because the applications of chromatography are so important and varied, commercialization of the technique is inevitable and certainly welcome. The contribution of instrument companies to the development of chromatography is noteworthy. Separation efficiencies of over a million theoretical plates and detection limits down to 10^{-22} g (a single molecule!) have become possible due to advancements in instrumentation. Several instrument companies have brought out recipe-book type manuals to enable the customers to use instruments and techniques without having to go into the basic principles. While this has served the user of chromatography in routine analytical laboratories, it has also hampered the development of the subject: the technique is seldom developed by the user, and as a consequence is not always tailor-made for a certain application. Solutions to newer separation problems require clear understanding of the basic principles as well as an overview of the options available. A prerequisite for this is easy access to structured information. The purpose of this series is to provide this information.

The subject of chromatography is too diverse and multi-faceted to be adequately presented by one or two authors in a single work. The structure of the series guarantees a comprehensive, well-balanced coverage of the theory, principles, methodology, instrumentation, practice and applications (both by class of compound and fields of application) of chromatography. However, each volume in this series can be read as a stand-alone book.

It is a pleasure to acknowledge the co-operation and support of the authors and editors of the individual volumes and the editorial staff of Harwood Academic Publishers.

B. Ravindranath, Ph.D.

Preface

This book has been written to demonstrate that chromatography is able to provide more information on solutes than just the confirmation of the presence of given species in individual samples and the assessment of their concentrations.

The specific information on solute properties can be gained by chemometric processing of retention data. Studies of that kind started some twenty years ago and early findings and methodology have been summarized in my previous book *Quantitative Structure-Chromatographic Retention Relationships*, John Wiley and Sons: New York (1987). In the last decade there has been great progress in this research area due to wide access to the advanced chemometric and molecular modeling software, as well as the development of modern chromatographic technology. Combined chromatographic–chemometric approaches to problem solving come into frequent use in chromatographic research and development, analytical chemistry, medicinal chemistry and drug design, environmental sciences, materials sciences, quality control, *etc.* Quantitative structure-retention relations (QSRR) also serve to test and improve chemometric methodology.

Several early chromatographic–chemometric findings and procedures have been confirmed and established in long term practice. However, some appeared to be incorrect or of limited value. This book is to account for changes within QSRR and give the reader exhaustive information regarding the most actual state of the art.

I did all my best to complete the accessible references. Some have not been quoted in this book because of methodological and interpretation errors. However, among hundreds of publications some works deserving attention could have been overlooked. For this I apologize to the authors.

The main aim of the book is to prove that a chemometric approach to chromatographic data produces useful, however probabilistic, information which is otherwise not easily accessible. Even if lacking precision, such information is of practical predictive value as opposed to the information provided by classical thermodynamics typified by the 'knowledge acquired after the fact'. The examples given show how to employ the chemometrically processed chromatographic parameters in order to predict the retention of new non-analysed solute, to elucidate molecular mechanisms of the processes determining retention in specific chromatographic systems, to identify the most informative structural descriptors of solutes, to evaluate the complex physicochemical properties within a series of solutes, and to predict relative differences in bioactivity of the chromatographed agents. I believe that after having read the book, chromatographers may find more efficient ways of exploiting their knowledge and skills and non-chromatographers may find chromatography a unique tool to help to solve their tasks.

Writing this book was partially supported by the Komitet Badań Naukowych, Warsaw, Poland. Thanks are due to Mrs. Janina Kozłów for her help in the typing of the manuscript.

To my Mother
For fifty years of care

To Anna, my wife
For twenty five years of great marriage

Also to her and to my sons Tomasz and Michał,
and to my daughter-in-law Justyna
For making my role of father unusually easy and highly rewarding

1 Introduction

Analytical chemists probably agree that in general, it is easier to synthesize a compound with a definite chemical structure than with a certain required property. Usually, reaction pathways can be deduced for standard chemical structures, whereas predicting properties of the compounds resulting from the reactions is still a matter of a scientific guess [1].

Let us imagine (as we often do) molecules as sets of balls (atoms) connected by stronger and weaker springs (bonds and interatomic interactions) colliding with another ball and spring system. Certainly, the crackings and fusions leading to a new entity can easily be understood. The reactivity seems to emerge as an innate feature of a molecule. This is not the case, however, with what is understood as physicochemical or biological (pharmacological) property of a chemical compound. Such a property depends not only on the compound's internal structure but also on the environment in which the compound is actually placed. From the point of view of properties the surrounding (environment) is by no means inert with regard to the molecules placed in it. The reactions with molecular environment, and not the molecular structure of compounds alone justify calling them drugs, toxins, hormones, feromones, odorants, pesticides, herbicides, environmental pollutants, cosmetics, detergents, impregnants, conductors, building materials, *etc.* Unlike chemical reactions, the interactions of molecules which form the environment with the molecules (or their fragments) placed in the environment cause neither the breaking of existing bonds nor the formation of new bonds.

It is a well known fact that in assigning properties to chemical structures, the classical thermodynamic approach appears inappropriate in practice. The reason is that thermodynamic properties of a given system are bulk properties reflecting only the net interactive effects in that system [2]. The magnitude of thermodynamic parameters represents the combination of individual interactions which may take place at the molecular or submolecular level. Thus, thermodynamic characteristics of chemical systems provides information of a physical rather than a chemical nature. As expressed by Prausnitz [3]: "Classical thermodynamics is revered, honored and admired, but in practice it is inadequate".

There is an approach to chemical problems which lacks the rigor of thermodynamics but can provide otherwise inaccessible information. This *extrathermodynamic* approach combines detailed models of physicochemical processes with certain concepts of thermodynamics. Extrathermodynamic in character are linear free-energy relationships (LFERs), commonly acknowledged by chemists [4]. It is believed that demonstration of LFERs suggests the presence of a real connection between correlated quantities, and that the nature of this connection can be identified [5]. In other words, one assumes that correlations among specific quantities are attributable to some unknown physicochemical relationships. The correlations (derived statistically) encourage the attempts to identify the real relationships behind them. Assuming LFERs one can treat a physicochemical-property-measuring-system (*e.g.*, chromatographic column) as a "free-energy transducer" translating differences in chemical potentials of solutes, arising from differences in structure, into quantitative differences in the property (*e.g.*, chromatographic retention).

Separation scientists use conventional relative measures of solute migration in individual chromatographic techniques. These are well known Kováts indices in gas chromatography, logarithms of capacity factors in column liquid chromatography, R_M values in thin-layer chromatography and electrophoretic mobilities in capillary zone electrophoresis. These "retention" parameters are commonly assumed to depend on the free-energy change associated with the solute distribution between the stationary and the mobile zones of the separation system. The existence of LFERs is normally proved statistically. To get statistically significant relationships and to avoid chance correlations one needs relatively large series of reliable parameters. Chromatography is a unique method which can readily yield a great amount of quantitatively comparable, precise and reproducible retention data for large sets of structurally diversified solutes. These data can be mutually related because all are obtained at the same experimental conditions (or are standardized by simple extrapolation/interpolation procedures).

LFERs in chromatography were reported first in 1950 by Martin [6]. Martin postulated that a substituent changes solute retention parameter by a factor dependent on the nature of the substituent and on both the mobile and stationary phases. In 1963 Green *et al.* [7] found that substituent increments to the thin-layer chromatographic parameter, R_M, are additive for a number of benzenoid compounds. In 1977 three publications appeared on what is now termed quantitative structure-retention relationships (QSRRs) [8–10]. These papers initiated extensive studies on the subject. A comprehensive monograph on QSRR published in 1987 [11] considered several hundreds original publications. By now this number can be estimated about 1000.

First QSRR papers resulted from applying the methodology used for studies of quantitative structure-biological activity relationships (QSARs) – the so-called Hansch approach [12] – to the analysis of chromatographic data. Latter on various modern chemometric methods of data processing have been introduced to QSRR. Probably, the chromatographic retention data belong to that type of physicochemical data which area most widely processed chemometrically. It can be stated without exageration that chemometrics owes to a great extent its present well established position among chemical disciplines to QSRRs.

Methology and goals of QSRR studies is schematically presented in Fig. 1.1. To undertake the study one needs a set of quantitatively comparable retention data for a sufficiently large series of solutes and a set of structural parameters of solutes (structural descriptors). Through the use of computerized statistical chemometric techniques, retention parameters are characterized in terms of various solute descriptors and their combinations and transformations (*e.g.*, squares, square roots, reciprocals, *etc.*). If statistically significant and physically meaningful QSRRs are obtained then they can be applied to:

i) predict retention for a new solute
ii) identify the most informative (regarding properties) structural descriptors
iii) gain insight into molecular (submolecular) mechanism of separation operating in a given chromatographic system
iv) evaluate other than chromatographic, complex physicochemical properties of solutes, *e.g.*, their hydrophobicity
v) predict relative biological activities within a set of drugs and other xenobiotics as well as material properties of a family of chemicals.

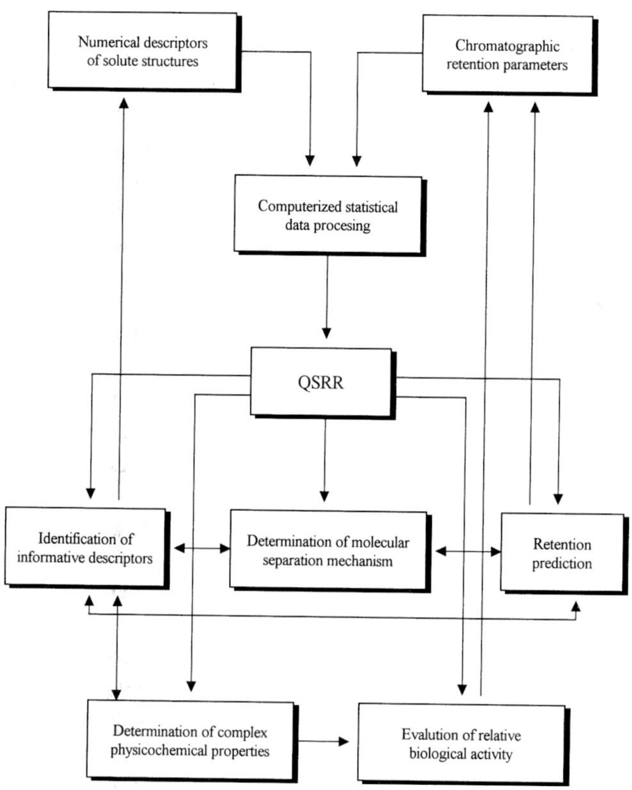

Figure 1.1. Quantitative structure-retention relationships (QSRR) studies: methodology and goals. (Adapted with permission from Kaliszan, R., *Anal. Chem.* 64, 619A, 1992. Copyright 1992 American Chemical Society.)

To obtain reliable QSRR appropriate input data must be provided and a stringent statistical analysis must be carried out. Chromatography can provide valuable input data. In a chromatographic process all conditions may be kept constant. Thus, solute structure becomes the single independent variable in the system.

QSRR studies are especially attractive from the general chemometric point of view. The reason is that QSRRs appear to be the best testing of the applicability of individual structural parameters for property description. The skill and knowledge gained from QSRR analysis may be applicable to other chemometric studies.

Undoubtedly, numerous QSRR studies in the literature deserve the interest and attention of analytical, physical, environmental and medicinal chemists. One has to be cognizant, however, of the fact that not every published QSRR equation provides reliable information. Some equations are statistically invalid, and sometimes statistically valid correlations are developed for chemically invalid principles. Nonetheless, many QSRR studies deserve the interest of chemist and other specialists.

References

1. Kaliszan, R., *Anal. Chem.* **64**, 619A (1992).
2. Kaliszan, R., in: "*High Performance Liquid Chromatography*", Brown, P.R. and Hartwick, R.A. (Eds.), Wiley, New York, 1989; Chapter 14, p. 563.
3. Prausnitz, J.M., *Science* **205**, 759 (1979).
4. Leffler, J.E. and Grunwald, E., "*Rates and Equilibria of Organic Reactions*", Wiley, New York, 1963.
5. Reichardt, C., "*Solvent Effects in Organic Chemistry*", Verlag Chemie, Weinheim, 1979; p. 227.
6. Martin, A.J.P., *Annu. Rev. Biochem.* **19**, 517 (1950).
7. Green, J., Marcinkiewicz, S. and McHale, D., *J. Chromatogr.* **10**, 158 (1963).
8. Kaliszan, R. and Foks, H., *Chromatographia* **10**, 346 (1977).
9. Kaliszan, R., *Chromatographia* **10**, 529 (1977).
10. Michotte, I. and Massart, D.L., *J. Pharm. Sci.* **66**, 1630 (1977).
11. Kaliszan, R., "*Quantitative Structure-Chromatographic Retention Relationships*", Wiley, New York, 1987.
12. Hansch, C. and Fujita, T., *J. Am. Chem. Soc.* **86**, 1616 (1964).

2 Basis of Chromatographic Separations

Chromatography is one of about 20 basic techniques of separation. There are numerous variants of chromatography named after typical physicochemical or operation features. Major variants of chromatography are collected in Table 2.1 after Giddings [1].

Obviously, the chromatographic separation is a result of the differential migration of components. Like in other separation systems, also in chromatography the driving force for separative displacement is equilibrium which favours the desired displacement.

There are two main classes of equilibria in isolated systems. Mechanical equilibrium which defines the resting place of macroscopic bodies, and molecular equilibrium which defines the spatial distribution of molecules (and colloids) at equilibrium [2]. Contrary to mechanical equilibrium, considerations of molecular equilibrium are complicated by entropy.

From the point of view of fundamentals of separations the thermodynamics of equilibrium between phases and equilibrium in external force fields is important. These two forms of equilibrium underlie the primary driving forces in most separation systems. Excellent treatment of the relevant thermodynamics of separation systems was provided by Giddings [2]. Below the basic ideas are given as extracted from this treatment.

The internal energy increment $\mathrm{d}E$ for a closed system (in which such changes as expansion, contraction, internal diffusion, chemical interaction, heating, *etc.*, may occur) is according to the first law of thermodynamics:

$$\mathrm{d}E = q + w, \tag{2.1}$$

where q is the increment of added heat and w is the increment of work done on the system. Assuming that only pressure-volume work is involved one gets $w = -p\,\mathrm{d}V$. The negative sign is because positive work is done on the system only when there is contraction, *i.e.*, when $\mathrm{d}V$ is negative. According to the second law of thermodynamics q is related to entropy S and temperature T by inequalities: $\mathrm{d}S \geq q/T$ or $T\,\mathrm{d}S \geq q$.

Table 2.1 Major variants of chromatography [1].

No.	*Variable of chromatographic system*	*Variations (techniques) of chromatography*
1.	Mobile phase	Gas
		Liquid { Normal phase; Reversed phase }
		Supercritical fluid
2.	Mechanism of retention	Adsorption Partition Bonded phase Ion exchange Exclusion Hydrophobic
3.	Incoming solute profile	Zonal Frontal
4.	Isotherm linearity	Linear
		Nonlinear { Displacement; Zonal }
5.	Column or planar	Column { Packed; Capillary }
		Planar { Thin-layer; Paper }
6.	Gradients	Nongradient
		Gradient { Gradient elution; Programmed temperature; Chromathermography; Programmed flow }
7.	Dimensionality	One-dimensional Two-dimensional (Continuous; Discrete) Multidimensional
8.	Physical scale	Microbore Standard Preparative scale

Thus Eq. (2.1) can be rewritten in the following form:

$$dE \leq T\,dS - p\,dV. \tag{2.2}$$

As known the Gibbs free energy, G, relates to enthalpy, H, and entropy, S:

$$G = H - TS = E + pV - TS. \tag{2.3}$$

After direct differentiation one gets

$$dG = dE + p\,dV + V\,dp - T\,dS - S\,dT. \tag{2.4}$$

Now one can substitute Eq. (2.2) for dE in Eq. (2.4) receiving

$$\mathrm{d}G \leq -S\,\mathrm{d}T + V\,\mathrm{d}p. \tag{2.5}$$

It comes out from Eq. (2.5) that all natural processes occuring at constant T and p must have d$G \leq 0$, while for any change at equilibrium d$G = 0$. This means that equilibrium at constant T and p is characterized by a minimum in G. Spontaneous processes, *e.g.*, diffusion are of course accompanied by d$G < 0$.

In open systems the expression describing dG at equilibrium in closed systems must be modified. An open system is defined as such in which can take place all the changes allowed for a closed system and which can in addition lose and gain matter across system boundaries. For open systems the equation

$$\mathrm{d}G = -S\,\mathrm{d}T + V\,\mathrm{d}p \tag{2.6}$$

has to be modified to account for small amounts of free energy, G, taken in and out of the system by the matter crossing its boundaries. If dn_i moles of component i enter the system whereas T and p do not change and no other component j crosses in or out, then G will change by a small increment proportional to dn_i:

$$\mathrm{d}G = (\delta G/\delta n_i)_{T,p,n_j}\,\mathrm{d}n_i. \tag{2.7}$$

This equation demonstrates that if the other factors are constant the magnitude of the increment depends on the rate of change of G with regards to n_i. The rate of change, or partial derivative is usually denoted by μ_i and called the chemical potential

$$\mu_i = (\partial G/\partial n_i)_{T\ p,n_j}. \tag{2.8}$$

Chemical potential is the amount of energy, G, brought into a system per mole of added constituent, i, at constant T and p. Applying the symbol μ_i the Eq. (2.7) can be rewritten as follows:

$$\mathrm{d}G = \mu_i\,\mathrm{d}n_i. \tag{2.9}$$

Analogously the addition of dn_j moles of j to the system yields d$G = \mu_j\,\mathrm{d}n_j$. For the transfer of any number of components, dG is a sum of individual contributions:

$$\mathrm{d}G = \sum \mu_i\,\mathrm{d}n_i. \tag{2.10}$$

Let us consider the case when T and p are allowed to vary. In such a situation the increment which describes their contribution must be added which is identical to that for a closed system Eq. (2.6). All of these contributions taken jointly result in a general equilibrium equation for an open system subject to any combination of incremental changes:

$$\mathrm{d}G = -S\,\mathrm{d}T + V\mathrm{d}p + \sum \mu_i\,\mathrm{d}n_i. \tag{2.11}$$

In Eq. (2.11) the summation extends over all components entering (dn_i positive) or leaving (dn_i negative) the system.

The μ_i value for a component i (*e.g.*, solute) in a given open system (*e.g.*, chromatographic phase) depends on two factors. One factor is the intrinsic thermodynamic "affinity" of the component to the system. The second is the dilution of the component. Dilution affects μ_i through entropy. In analytical separations one normally deals with high dilutions (low solute concentrations). Then the "affinity" and "dilution" contributions to chemical potential are represented by μ_i^0 and c_i in the following equation:

$$\mu_i = \mu_i^0 + RT \ln c_i. \tag{2.12}$$

In Eq. (2.12) R means the gas constant, T is the temperature, c_i is the concentration of component i (moles or mass per unit volume) and μ_i^0 is the so-called standard-state chemical potential. The μ_i^0 is the chemical potential in a hypothetical standard state with component i at unit concentration but with each molecule of i surrounded by solvent as found at infinite dilution. The value of μ_i^0 depends strongly on the energy and to some extent on the entropy of component-environment (solute–solvent) interactions. It is thus a function of chemical structure of interacting entities. The problem is, however, how μ_i^0 is to be calculated. Unfortunately, thermodynamics is helpless in that respect [2,3].

The term $RT \ln c_i$ in Eq. (2.12) accounts for a contribution to chemical potential due to the entropy associated with the dilution and concentration of components. Equilibrium in two open systems α and β implies equality $\mu_i^\beta = \mu_i^\alpha$. Substituting Eq. (2.12) to this equality and solving the resulting expression for the ratio of the two concentrations c_i^α and c_i^β one gets the equilibrium equation:

$$(c_i^\beta / c_i^\alpha)_{\text{eq}} = \exp(-\Delta\mu_i^0 / RT), \tag{2.13}$$

where $\Delta\mu_i^0 = \mu_i^{0\beta} - \mu_i^{0\alpha}$. The equilibrium ratio $(c_i^\beta / c_i^\alpha)_{\text{eq}}$ is the partition coefficient or distribution coefficient K:

$$K = \exp(-\Delta\mu_i^0 / RT). \tag{2.14}$$

Because at equilibrium the μ_i in Eq. (2.12) is equal for component i in two partitioning systems ($\mu_i^\beta = \mu_i^\alpha$), the system (phase) with lower μ_i^0 must compensate with a larger $RT \ln c_i$ term, *i.e.*, with higher c_i. Thus, if $\Delta\mu_i^0$ is positive the μ_i^0 in system α (termed $\mu_i^{0\alpha}$) is lower than in system β. Therefore, the concentration in α is higher than in β.

It is clear now that separative equilibria involve adjustments in concentration to satisfy the criterion $\mu_i^\beta = \mu_i^\alpha$. Different intermolecular interactions give different $\Delta\mu^0$ values for individual components. Thus unequal concentration distributions arise which ultimately lead to separation.

In some separation systems equilibrium distribution of chemical components is affected by external fields, *e.g.*, electromagnetic, gravitational. A field imparts to a component (molecule, particle) a certain potential energy varying with position. This potential energy is an additive input to the Gibbs free energy, G. The potential energy for one mole becomes an additive contribution to chemical potential designated as μ_i^{ext} for component i. This term is different from the chemical potential having internal origins μ_i^{int} (previously denoted μ_i). Thus, in case of operating external fields Eq. (2.11) can be modified as follows:

$$dG = -S\,dT + V\,dp + \sum(\mu_i^{\text{int}} + \mu_i^{\text{ext}})\,dn_i. \tag{2.15}$$

Considering constant T and p one gets

$$dG = \sum(\mu_i^{\text{int}} + \mu_i^{\text{ext}})\,dn_i. \tag{2.16}$$

After transferring dn_i moles of component i from region α to region β at equilibrium the following equation holds:

$$dG = 0 = (\mu_i^{\text{int},\beta} + \mu_i^{\text{ext},\beta} - \mu_i^{\text{int},\alpha} - \mu_i^{\text{ext},\alpha})\,dn_i = (\Delta\mu_i^{\text{ext}} + \Delta\mu_i^{\text{int}})\,dn_i. \tag{2.17}$$

Now, one can substitute Eq. (2.12) for each μ_i^{int}. In effect Eq. (2.17) yields

$$-(\Delta\mu_i^{\text{ext}} + \Delta\mu_i^0) = RT\ln(c_i^\beta/c_i^\alpha)_{\text{eq}}. \tag{2.18}$$

After rearrangement it gives

$$K = (c_i^\beta/c_i^\alpha)_{\text{eq}} = \exp[(-\Delta\mu_i^0 - \Delta\mu_i^{\text{ext}})/RT]. \tag{2.19}$$

Eq. (2.19) is similar to Eq. (2.14), except for the addition of the external field term $\Delta\mu_i^{\text{ext}}$.

The thermodynamics outlined above is universally applicable to all dilute systems. However, it is lacking specifics. As pointed out by Giddings [2] the $\Delta\mu_i^0$ cannot be related rigorously by any means to molecular parameters. There are approximate theories available for description of the distribution of solute between phases, *e.g.*, a lattice model of Martire and co-workers [4,5]. However, their application requires usually a knowledge of number of not easily attainable parameters. Giddings [2] concludes that "Because pure theory is impractical, progress in understanding and describing molecular equilibrium between phases requires a combination of careful experimental measurements and correlations by means of empirical equations and approximate theories". Such an approach is the subject of this book.

References

1. Giddings, J.C., "*Unified Separation Science*", Wiley, New York, 1991, Chapter 10.
2. Giddings, J.C., "*Unified Separation Science*", Wiley, New York, 1991, Chapter 2.
3. Scott, R.P.W., in: "*High Performance Liquid Chromatography*", Brown, P.R. and Hartwick, R.A. (Eds.) Wiley, New York, 1988, Chapter 2.
4. Martire, D.E. and Boehm, R.E., *J. Phys. Chem.* **91**, 2433 (1987).
5. Martire, D.E., *J. Chromatogr.* **452**, 17 (1988).

3 Intermolecular Interactions Controlling Chromatographic Separations

3.1. Thermodynamic Basis

The standard-state chemical potential μ_i^0 is the gain in the Gibbs free energy per mole of added component i under standard conditions. The actual chemical potential value μ_i is composed of both enthalpy and entropy terms. In case of a two-phase system the parameter $\Delta\mu_i^0$ is the difference in two μ_i^0 values and can be expressed as follows [1]:

$$\Delta\mu_i^0 = \Delta\hat{H}_i^0 - T\Delta\hat{S}_i^0, \tag{3.1}$$

where symbols $\hat{H}^0$ and $\hat{S}^0$ represent the partial molar enthalpy and entropy under standard conditions, respectively.

Generally, in separation systems involving partitioning of solutes between phases the absolute value of $|\Delta\hat{H}^0|$ is much higher than $|T\Delta\hat{S}^0|$. In other words, the equilibrium is mostly determined by enthalpy factors. As pointed out by Giddings [1], the increment $\Delta\hat{H}_i^0$ is most often controlled by intermolecular interactions between component i and the two phases it occupies. If intermolecular attractions in phase β are stronger (lower enthalpy) than in phase α then the term $\Delta\hat{H}_i^0$ is negative ($\Delta\hat{H}_i^0 = \hat{H}_i^{0\beta} - \hat{H}_i^{0\alpha}$). In such a situation $\Delta\mu_i^0$ tends to be negative and the distribution constants K greater than 1 favouring phase β. This is as suggested by chemical intuition: the more strongly a solute interacts with a phase the more of it will partition into this phase.

3.2. Intermolecular Interactions

The question arises now what is the nature of the intermolecular interactions governing the chromatographic separation. Certainly these are not the interactions causing definite chemical alterations of the solute molecules. In other words, these are not the protonation, oxidation, reduction, complex formation nor other stoichiometric processes [2]. Only in ion-exchange chromatography, where the separation determining forces are ionic in nature, can discrete chemical alterations be said to occur. In other chromatographic techniques and modes only the forces that can occur between closed-shell molecules are involved [3]. Covalent binding formation leads to formation of a new chemical entity and is not normally involved in chromatographic separations. In Table 3.1 known in chemistry binding and intermolecular interaction types and energies are given after Albert [4] and Seydel and Schaper [5].

3.2.1. Ion–Dipole Interactions

Of chromatographic importance can be the ion–dipole interactions. In fact, Coulomb forces between ions and electrically neutral molecules possesing permanent dipole moments do not belong to the intermolecular forces in the narrower sense [6]. These coulombic forces, however, may play an important role in some chromatographic processes.

When placed in the electric field produced by an ion, the dipole will orient itself so that the end with charge opposite to that of the ion will be directed toward the ion, and the other repulsive end directed away. The potential energy of an ion–dipole interaction $E_{i\text{-}d}$ is described by the following equation:

$$E_{i\text{-}d} = -\hat{W}^2 Z \mu (\cos \alpha')/\varepsilon r^2, \qquad (3.2)$$

Table 3.1 Binding types and binding energies with example systems. Modified after Albert [4] and Seydel and Schaper [5].

Kind of Interaction	*Energy of Interaction (kJ/mol)*	*Example System*
Covalent Binding	−(170–460)	CH_3–OH
Strong Ionic Binding	−40	HRN–H - - O; $H^{\oplus}$ - - - $^{\ominus}$O; C–R′
Ionic Binding	−20	$R_4N^{\oplus}$- - - $Cl^{\ominus}$
Ion–Dipole Interactions	−(4–17)	$R_4N^{\oplus}$- - -:NR_3
Dipole–Dipole Interactions	−(4–17)	O$=$$C^{\delta+}$- - -:$NR_3$
Hydrogen Binding Interactions	−(4–17)	—OH - - - O$=$
Electron Pair Donor–Electron Pair Acceptor Interactions	−(4–17)	—OH - - - C$=$C
Hydrophobic Interactions	−4	R CH_2 CH_2 R ⋮ R CH_2 CH_2 R
Dispersive Interactions	−(2–4)	C - - - - C

where the coefficient $\hat{W}$ depends on the unit system applied. For the SI unit system $\hat{W} = 1/(4\pi\varepsilon_0)$, where $\varepsilon_0 = 8.854 \times 10^{-12}$ F/m is the electric permittivity of vacuum. In Eq. (3.2) Z is the charge of the ion, μ is the dipole moment of the neutral molecule, r is the distance from the ion to the center of the dipole, α' is the dipole angle relative to the line joining the ion and the center of the dipole and ε is the relative electric permittivity of the medium (eluent). One can note that the higher the relative electric permittivity ε of the medium, the lower is the energy of attractive ion–dipole interactions. In such a situation a decrease of eluent permittivity by adding ethanol ($\varepsilon = 27$) to water ($\varepsilon = 81$) should increase strength of electrostatic interactions.

3.2.2. Dipole–Dipole Interactions

The long-distance forces resulting from the electric field produced by the molecules (nonionized) are called Van der Waals interactions. These interactions are usually classified into three groups: orientation, inductive and dispersive interactions.

The potential energy $E_{d\text{-}d}$ of orientation interactions (Keesom effect) between molecules of dipole moments μ_1 and μ_2 is

$$E_{d\text{-}d} = -2\hat{W}^2\mu_1^2\mu_2^2/(3kT\varepsilon r^6). \tag{3.3}$$

3.2.3. Dipole–Induced Dipole Interactions

Electric dipole in one molecule can induce a dipole moment in a neighbouring molecule. The induced dipole moment always lies in the direction of the inducing dipole. Thus, attraction always takes place between the two molecules which is independent of temperature. The higher the polarizability α of the apolar molecule, the larger will be the induced moment. The potential energy $E_{d\text{-}id}$ of the dipole–induced dipole interactions (Debye effect) is given by the following equation:

$$E_{d\text{-}d} = -\hat{W}^2(\alpha_1\mu_2^2 + \alpha_2\mu_1^2)/\varepsilon r^6. \tag{3.4}$$

The dipole–induced dipole interactions are not additive and do not depend on temperature.

3.2.4. Instantaneous Dipole–Induced Dipole Interactions

Classical electrodynamics does not explain why two electrically neutral molecules possessing no permanent dipole moment attract one another at close distances. However, even noble gases may exist in the liquid state, which means that their atoms, characterized by spherically symmetrical charge distribution, interact.

In atoms and molecules possessing no permanent dipole moment, the continuous electron density fluctuations result, at any instant, in a small dipole moment. This instantaneous dipole can fluctuatingly polarize the electron system of the neighbouring atoms or molecules. The approximate equation for the calculation of the potential energy E_d of such dispersion interactions (London effect) is

$$E_d = -2\hat{W}_2 I_1 I_2 \alpha_1 \alpha_2/[3(I_1 + I_2)\varepsilon r^6], \tag{3.5}$$

where I_1 and I_2 are the ionization energies of the interacting molecules 1 and 2. In Eq. (3.5) molecular polarizabilities α_1 and α_2 of the interacting molecules are given as scalar quantities. Accordingly, the dispersion interactions cannot be saturated and are additive. This type of intermolecular interactions is the most catholic of the three kinds of Van der Waals interactions.

Examination of Eqs. (3.3)–(3.5) leads to conclusion that Van der Waals interactions increase with pressure as the intermolecular distances r decrease.

Van der Waals interactions are commonly classified as chemically nonspecific. There are also intermolecular interactions in chromatography which are assigned to characteristic structural features of the molecules involved. If Van der Waals interactions are more "physical" in nature, these structurally specific interactions have a more "chemical" character.

3.2.5. Hydrogen Bonding Interactions

When a covalenthy bound hydrogen atom forms a second bond to another atom, the second bond is referred to as a hydrogen bond. The proton donor structural features of molecules are such functional groups like –O–H, –S–H, Cl–H, Br–H. The most important hydrogen bond acceptors are the oxygen atoms in alcohols, ethers and carbonyl compounds, halogen atoms and ions, nitrogen atoms in amines and the π-electron systems of aromatic compounds.

3.2.6. Electron Pair Donor–Electron Pair Acceptor Interactions

The second important type of "chemical" intermolecular interactions are the so-called electron pair donor–electron pair acceptor (EPD–EPA) interactions (charge transfer interactions). Whereas in a typical chemical bond each atom supplies one electron to the bond, in EPD–EPA bonding one molecule supplies the pair of electrons while the second molecule provides the vacant molecular orbital. The formation of such an additional bonding interaction between two valency-saturated molecules is possible only if an occupied molecular orbital of sufficiently high energy is present in the electron pair donor molecule and a sufficiently low energy occupied orbital is available in the electron pair acceptor molecule. The energetically highest orbitals are the lone pairs of the n electrons (*e.g.*, amines, ethers, sulphoxides, phosphines), the electron pair of δ bond (*e.g.*, cyclopropane, halogen derivatives) and the pair of π electrons of unsaturated and aromatic compounds (*e.g.*, polyaromatic compounds). The electron pair acceptor compounds are those possessing a vacant valency orbital of a metal atom (*e.g.*, organometalic compounds), a nonbonding σ orbital (*e.g.*, halogen molecules) and compounds having a system of π bonds (*e.g.*, aromatic polynitro compounds, tetracyanoethylene).

3.2.7. Hydrophobic Interactions

The distribution of solutes between the mobile and the stationary phases in liquid chromatography, especially in the so-called reversed-phase mode, is often interpreted in terms of hydrophobic (solvophobic) interactions [7]. The phenomenon of hydrophobic interactions is the subject of controversy [8]. However complex, the hydrophobic interactions are the net effect of the above discussed well-known physicochemical interactions which determine the state of all matter.

3.3. Chromatographic Retention in view of Fundamental Intermolecular Interactions

The intermolecular interactions taking place during a chromatographic separation process appear to be quite simple. However, the details of the mechanism of retention are not fully understood. Obviously, one can assume that the distribution of a solute between a mobile zone and stationary zone during the chromatographic process results from the forces which operate between solute molecules and the molecules present in each zone.

Let us consider, for example, the reversed-phase liquid chromatography (RP LC) on chemically bonded hydrocarbon silica stationary phase. In this case one has to consider the following interacting entities: the solute, the hydrocarbon bonded to silica, the mobile phase components preferentially adsorbed on the stationary phase material, the free silanol groups of the silica support and the components of the eluent. The situation appears quite complex and no satisfactory model exists which would permit quantitative prediction of retention. Nonetheless, a rational explanation of the observed retention changes in terms of intermolecular interactions appears possible.

A typical feature of RP LC is an increase of logarithm of capacity factor ($\log k'$) of a solute with an increasing number of carbon atoms for homologous series. Considering first the interactions of solutes with the hydrocarbonaceous moiety of stationary phase one identifies the dispersive forces as differentiating the homologues. One can assume the input to separation due to the orientation interactions of homologues as negligible because the polarity (dipole moment) of the hydrocarbon part of the molecules is practically zero. Also the dipole–induced dipole interactions should be similar for all homologues because the dipole moments within homologous series are similar. Of course, the magnitude of dispersion interactions increases with increasing polarizability of the solutes, which actually reflects their molecular size (bulkiness). In the case of solute-polar eluent interactions of homologous solutes the orientation forces and the inductive forces are undoubtedly stronger than the solute-nonpolar hydrocarbon of stationary phase interactions. Certainly, the orientation interactions with eluent are similar for all homologues due to the similarity of their dipole moments.

The attraction of homologous solutes by a mobile phase resulting from the dispersive and inductive interactions is affected by the solute polarizability. However, because the dispersive interactions are usually prevailing among intermolecular interactions and the polarizability of hydrocarbon, *e.g.*, octadecyl chains of the stationary phase, is greater than the polarizability of the smaller molecules of typical eluents used in RP LC, the net effect of Van der Waals interactions will be the increased retention of larger homologues. Of course, the interactions of solutes with the components of a chromatographic system can be further complicated by the "chemical" type interactions previously mentioned.

It is an experimental fact that for a given solute the RP LC retention normally decreases with increasing content of organic modifier in binary aqueous eluent. One reason for this may be that due to slightly higher polarizability of the organic/water eluent (as compared to water) the dispersive interactions of a solute with the mobile phase increase with the content of the organic component. This should result in some decrease of retention. The other reason for the observed retention decrease with increasing organic modifier content in the mobile phase seems to be the increase of the strength of all electrostatic interactions between solutes and the eluent resulting from a decrease of the relative electric permittivity of the aqueous eluent after diluting water with organic solvent Eqs. (3.2)–(3.5).

Another characteristic RP LC phenomenon is the increase of retention with increasing ionic strength of eluent. Again, this can be rationalized based on fundamental intermolecular interactions. Whereas the direct interactions of a solute with stationary phase should not be much affected by changes in the ionic strength of the eluent, the attraction of solute molecules by mobile phase molecules due to the Van der Waals interactions would be diminished with increasing concentration of ionic species. This explains why increase of the ionic strength of aqueous eluents leads to increasing retention in RP LC.

The effect of eluent pH on the retention of weak acids and bases in RP LC on hydrocarbon-bonded silica also deserves discussing. One can assume that the dispersive interactions of solute ions with both phases do not differ significantly from these interactions for nonionized solute molecules. However, in the case of solute ions the ion–dipole interactions become dominating. Such attractive interactions are of practical importance between solute ions and polar molecules of RP LC eluent as opposed to nonpolar hydrocarbon moieties of stationary phase. This is true especially since the ion–dipole long-distance interactions are stronger than regular short-range Van der Waals interactions and decrease with the second power of the distance between the interacting species. From the above discussion the observed in RP LC decreased retention of solutes with their increasing degree of ionization appears reasonable.

It is often reported that the retention of the completely ionized acids in RP LC on hydrocarbon-bonded silica is lower than the retention of ionized bases of similar molecular size (bulkiness). This may be explained in terms of interactions with free silanol sites on the surface of the stationary phase. The free silanols produce some exclusion effects towards anions whereas the retention of cations is increased. This type of electrostatic (coulombic) interactions is reduced at higher ionic strength of the eluent. The effect becomes negligible at ionic strength above 0.1 [9].

A characteristic feature of RP LC is the observed decrease of solute retention with increasing temperature. This fact can also be interpreted in terms of the theory of fundamental intermolecular interactions. It is known that with increasing temperature the dielectric constant of water (the component of RP LC eluents) decreases significantly: from 88.0 at 0°C to 63.7 at 70°C. Assuming the dispersive interactions of a solute with the hydrocarbonaceous stationary phase to be basically unaffected by temperature changes, one can expect decreasing at elevated temperature retention because the accompanying decrease in the dielectric constant increases affinity of the solutes to the eluent.

References

1. Giddings, J.C., "*Unified Separation Science*", Wiley, New York, 1991, Chapter 2; p. 24.
2. Kaliszan, R., *Crit. Rev. Anal. Chem.* **16**, 323 (1986).
3. Kaliszan, R., in: "*High Performance Liquid Chromatography*", Brown, P.R. and Hartwick, R.A. (Eds.), Wiley, New York, 1988, Chapter 14; p. 565.
4. Albert, A., "*Selective Toxicity*", *3rd ed.*, Wiley, New York, 1965.
5. Seydel, J.K. and Schaper, K.-J., "*Chemische Struktur und biologische Aktivität von Wirkstoffen*", Verlag Chemie, Weinheim, New York, 1979, p. 14.
6. Reichardt, C., "*Solvent Effects in Organic Chemistry*", Verlag Chemie, Weinheim, New York, 1979, p. 9.
7. Horváth, C., Melander, W. and Molnar, J., *J. Chromatogr.* **125**, 129 (1976).
8. Ben-Naim, A., "*Hydrophobic Interactions*", Plenum, New York, 1980.
9. Knox, J.H., Kaliszan, R. and Kennedy, G.J., *J. Chem. Soc., Faraday Symp.* **15**, 113 (1980).

4 Structural Descriptors of Solutes

4.1. Linear Free-Energy Relationships

As shown in the preceeding chapter, a rationalization of the observed chromatographic behaviour of solutes at specific separation conditions appears possible in terms of the fundamental intermolecular interactions. In spite of that, no strict quantitative retention model exists which would permit a precise prediction of the chromatographic parameters for individual solutes analysed at given conditions. The reason is that statistical mechanics, the science which should yield required information, is hampered by the "multibody" complexity of molecular interactions in condensed phases and by the inadequacy of quantum mechanics in providing accurate interaction potentials between molecules [1].

Because pure thermodynamics appears impractical in modeling chromatographic processes the approach called extrathermodynamic is applied to extract useful chemical information from retention data [2]. The term "extrathermodynamics" denotes that the science lies outside the formal structure of thermodynamics, although the approach resembles that of thermodynamics in that detailed microscopic mechanism does not need to be explicity identified during use [3]. Extrathermodynamic approaches are combinations of detailed models with the concepts of thermodynamics. As these approaches involve model building they lack the rigour of thermodynamics but can provide information not otherwise accessible. The commonly recognized manifestations of extrathermodynamic relationships are the linear free-energy relationships (LFER).

The LFER may be regarded as linear relationships between the logarithms of the rate or equilibria constants for one reaction series and those for a second reaction series subjected to the same variation in reactant structure or reaction conditions [4,5]. Plotting the logarithms of rate or equilibrium constant for one reaction series against the corresponding constants for a second series frequently gives a straight line. This expresses the following equation:

$$\log k_i^B = m \log k_i^A + n, \tag{4.1}$$

where k_i^A and k_i^B are rate or equilibrium constants of two reaction series A and B which are subject to the same changes in the structure of reactants or the surrounding medium. It is convenient to express LFER in terms of ratios of constants by referring all members of a reaction series to a reference member of the series. In such a situation the correlation in Eq. (4.1) can also be expressed in the following manner:

$$\log (k_i^B/k_0^B) = m \log (k_i^A/k_0^A), \tag{4.2}$$

where k_i^A and k_0^B are the constants for the reference substituent or the reference solvent.

In correlation studies of structure of solutes and their chromatographic behaviour the retention parameters are tacitly assumed to be linearly related to the free-energy change associated with the chromatographic distribution process. As a matter of fact, not all chromatographic data are suitable for correlation studies.

As is well known, free-energy changes, ΔG, are related to enthalpy, ΔH, and entropy, ΔS, changes by the Gibbs equation

$$\Delta G = \Delta H - T\Delta S, \tag{4.3}$$

where T is temperature. For LFER to be found between real and model system, changes in either entropy or enthalpy must be constant or the enthalpy changes must be linearly related to entropy changes [6]:

$$\Delta H = \beta \Delta S + \Delta G_\beta \quad \text{at } T = \beta. \tag{4.4}$$

When the so-called enthalpy–entropy compensation is observed with a family of compounds in a particular chemical transformation, the values of β and ΔG are invariant and β is called the compensation temperature. Then, as demonstrated by Melander *et al.* [6], plots of $\ln k'_T$ of various solutes measured at a given temperature T under different conditions against the corresponding enthalpy change are linear. From the slope of the compensation plot of $\ln k'_T$ against ΔH, the compensation temperature, β, may be obtained. It is recommended that the reference temperature T be near the harmonic mean of the experimental temperatures used for the evaluation of the enthalpies.

The majority of reported studies of quantitative structure-retention relationships (QSRR) concern chromatographic retention parameters as obtained in routine measurements. Enthalpy–entropy compensation is checked only occasionally and LFER are assumed *a priori.*

4.2. Structural Descriptors of Solutes based on Established Linear Free-Energy Relationships

After Martin [7] published his fundamental paper in 1950, followed by several publications by Green *et al.* [8], a great many reports have appeared in which substituent increments to chromatographic retention were determined. The contributions of the substituents to retention have often been found constant and additive for a given chromatographic system. In effect, reliable predictions of retention have been reported especially in the case of reversed-phase liquid chromatography or of gas chromatography on nonpolar stationary phases.

Especially important appears the paper by Tomlinson *et al.* [9]. These authors determined functional group contributions to reversed-phase high-performance liquid chromatographic (RP HPLC) retention for a set of methyl to decyl benzoates – both unsubstituted and *o*-, *m*-, *p*-OCH_3, $-NO_2$ and –Cl substituted. These group contributions were determined at conditions where enthalpy–entropy compensation was found. The group contribution τ was defined as follows:

$$\tau = \log r_{ij} = \log(k_i' \, k_j') = \log[(t_{Rj} - t_{R0})(t_{Ri} - t_{R0})], \tag{4.5}$$

where k' and t_R are the capacity ratio and retention times of solutes j and i which differ by a functional group. In Eq. (4.5) r is the chromatographic selectivity coefficient and t_{R0} is the migration time of a nonretained tracer. The τ values for different groups at different positions of substitution were found to depend linearly on the mobile phase composition as expressed by the eluent surface tension.

The most commonly recognized free-energy related parameters of substituents are Hammett constants. In 1937 Hammett formulated his equation for the calculation of substituent effects on reaction rates and chemical equilibria based on the dissociation of *m* or *p*-substituted derivatives of benzoic acid in water at 25°C:

$$\log(K_X/K_H) = \rho\sigma, \tag{4.6}$$

where K_X is the dissociation constant of the derivative possesing substituent X in the para or meta position, K_H is the constant for the unsubstituted acid, ρ is the reaction constant ($\rho = 1$ for the standard reaction by definition), σ_p or σ_m are the Hammett constants for substituents in an aromatic system located at the para or meta position, respectively. In the case of the ortho position, steric effects hamper unequivocal determination of the electronic constant for substituents.

As they are based on the extrathermodynamic assumption of LFER, the Hammett constants are additive. This means that for a multiple-substituted aromatic system the following relationship holds:

$$\log(K_R/K_H) = \rho\sum\sigma, \tag{4.7}$$

where K_R is the dissociation constant for the multiple-substituted derivative and $\sum\sigma$ is a sum of individual Hammett constants for substituents.

Good performance of classical Hammett constants in predicting differences in physicochemical properties among series of derivatives encouraged attempts to further differentiate the electronic substituent constants. Efforts were devoted to isolate and quantify more specific electronic effects like mesomeric, inductive, resonance and field effects of substituents.

Numerical values of Hammett constants and their various modifications can be found in compilations (*e.g.*, refs. [10–13]). Unfortunately, the available collections of data are incomplete as regards to many important substituents and the reliability of the reported constants is often disputable. This explains why the attempts to apply the electronic substituent constants in QSRR studies have met only limited success. Examples of QSRR analysis of GC, TLC and HPLC retention data involving the electronic substituent constants have critically been discussed elsewhere [2].

By analogy to the Hammett electronic substituent constant in early 1960s Hansch and co-workers introduced the substituent hydrophobic (lipophilic) constant π [14,15]. The π value for a given substituent is defined as the logarithm of the ratio of the n-octanol-water partition coefficient of a substituted to an unsubstituted derivative

$$\pi_X = \log(P_{R\text{-}X}/P_{R\text{-}H}), \tag{4.8}$$

where X denotes the substituent, R-H is the unsubstituted and R-X is the substituted compound for which the n-octanol-water partition coefficient P is experimentally determined. The standard π values are derived from benzene derivatives. The important feature of π constants is their additivity. Already in 1965 Iwasa *et al.* [16] reported significant correlation between π and the parameter ΔR_M from TLC. The parameter ΔR_M accounts for the effect of a substituent upon the R_M value of unsubstituted compound. Since that time numerous QSRR were published involving π (sum of π values for the substituents present in a molecule), especially regarding the so-called partition chromatographic data. Historical review of such QSRR studies can be found elsewhere [2,17,18].

Another classical linear free-energy related substituent constants defined from chemical reactivities is the Taft steric constant E_s [19]. Similarly to the electronic and hydrophobic substituent constants the E_s values have been compiled for more common substituents (*e.g.*, refs. [10–13]). There are only few examples of application of E_s in QSRR analysis of gas chromatographic retention data (*e.g.*, ref. [20]).

4.3. Solvatochromic Parameters

In late 1980s Carr and co-workers in studies on the nature of RP HPLC separations developed an approach based on the solvatochromic comparison method and the so-called linear solvation energy relationships (LSER) (*e.g.*, refs. [21–24]). The approach is based on a general equation describing logarithm of capacity factor, $\log k'$:

$$\begin{aligned}\log k' = \text{constant} + M(\delta_m^2 - \delta_s^2)V_x/100 + S(\pi_s^* - \pi_m^*)\pi_x^* \\ + A(\beta_s - \beta_m)\alpha_x + B(\alpha_s - \alpha_m)\beta_x,\end{aligned} \tag{4.9}$$

where the subscript x designates a solute property such as molar volume (V_x), polarizability/dipolarity (π_x^*), hydrogen bond acidity (α_x) and hydrogen bond basicity (β_x). Each solute property is multiplied by a term which represents the difference in complementary "solvent" properties of the mobile (subscript m) and the stationary (subscript s) phases. Thus, α_m and α_s are the abilities of the phases (bulk or bonded) to donate a hydrogen bond. These properties complement the solute's ability to accept a hydrogen bond (β_x). Similarly, δ_m^2 and δ_s^2, the squares of the Hildebrand solubility parameter or cohesive energies of the two phases, complement the solute intrinsic molar volume (V_x).

The parameters π^*, α and β are empirical. They were derived [25,26] from the analysis of the effects of solvents on the transition energies of the maximum absorption of the solvatochromic indicator compound [27]. Respective data are available for an increasing number of compounds [28–30]. Also the Hildebrand solubility parameters present in Eq. (4.9) have been compiled [31–33].

Although empirical, the solvatochromic parameters have the advantage of being fundamentally interpretable. That is, π^* is decoupled from compound hydrogen bonding processes and similarly compound parameters α and β are not related to compound dipolarity/polarizability [24].

4.4. Nonempirical Structural Descriptors

Under this term one can consider all the structural parameters which can be derived by computational chemistry based solely on the structural formula of a compound. As such can be treated the constitutive-additive parameters like molar refractivity, or $\log P$ calculated by fragmental methods, the quantum chemical indices calculated by semiempirical methods and the parameters derived from molecular graphs.

The number of structural descriptors which can be assigned to an individual solute is large enough to pose a statistical challenge. Especially, as it often is the case, if various transformations and combinations of particular descriptors are included. One of commercially available software systems can process 200 different structural descriptors [34].

Several computer programs are now available for calculation of the more commonly recognized structural descriptors. These programs were mostly developed for the analysis of quantitative structure-biological activity relationships (QSAR) but may equally well be used in studies of other structure-property relationships, including QSRR. Still, however, there is a space for individual researcher invention and imagination, especially when proposing specific *ad hoc* descriptors which are best suitable for individual tasks. The requirement is that the descriptors are unequivocally defined and identified, even if their physical meaning is occasionally obscure. Of course, the descriptors should characterize the three-dimensional molecular structure of the compounds and not their conventional representations by the planar chemical formulae.

4.4.1. CLOGP—the log *P* Calculated by Fragmental Methods

Large compilations of $\log P$ from the n-octanol-water partition system and observed additivity of π constants allowed Rekker [35] and later Hansch and Leo [10,15] for developing fragmental methods of calculating hydrophobicity of molecules from their structural formulae. The general equation to calculate $\log P$ in the fragmental systems is

$$\log P = \sum(\text{fragmental values}) + \sum(\text{correction terms}). \tag{4.10}$$

Example calculation based on fragmental and correction parameters according to Hansch and Leo [15] is illustrated in Fig. 4.1. Calculation of $\log P$ by fragmental methods requires experience (and intuition) to properly choose among the individual fragmental values and correction terms. Hand calculations seem often to be more reliable than those provided by commercially available software.

4.4.2. Refractivity

Molar refractivity and polarizability are the molecular "bulkiness" parameters which may be related to solute ability to undergo inductive and dispersive interactions with the components of chromatographic phases. For any given molecular formula refractivity may conveniently be calculated by summation of the atomic, the group or the bond-type increments. Independent of the method of calculation the refractivity values obtained are highly intercorrelated. The correlation between molar refractivity and molecular mass is moderate [19]. Reference data for calculation of refractivity may be found in literature [15,36–39].

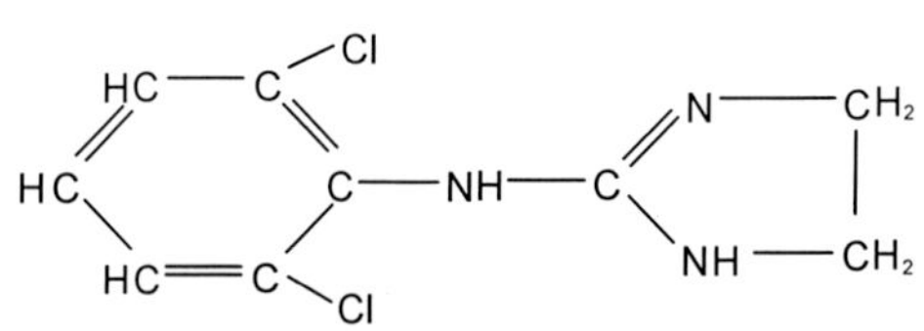

$f\,C_6H_3 = f\,C_6H_5 - 2f_H = 1.90 - 2 \times 0.23 = 1.44$
$2 \times f_{\underline{Cl}} = 2 \times 0.94 = 1.88$

$f_{\underline{NH}} = -1.03$
$(\overline{2-1})\,F_b = -0.12$
$F_g^{br} = -0.22$

$2 \times f_{CH2} = 2 \times 0.66 = 1.32$
$2 \times f_{NH} = 2 \times (-2.15) = -4.30$
$f_{CH} = 0.43$
$F_= \;= -0.55$
$(5-1)\,F_{\underline{b}} = 4 \times (-0.09) = -0.36$
$F_{\underline{p1}} + F_{\underline{p2}} = (-0.32 - 0.20) \times (-2.15 - 2.15) = 2.24$

Total sum of hydrophobicity fragmental constants, CLOGP = 0.73

The symbols used denote :
<u>NH</u>, <u>Cl</u> - NH, Cl aromatic;
NH - NH aliphatic;
$\underline{b}$ - single bond in ring;
b - single bond in chain;
= - double bond;
g^{br} - group branching;
$p_{\underline{1}(2)}$ - proximity effect in ring (2 polar fragments separated by 1 or 2 carbon atoms).

Figure 4.1. Example calculation of $\log P$ by the fragmental method of Hansch and Leo [10,15].

4.4.3. Polarizability

The most commonly used presently method of estimating compounds polarizabilities is based on the additivity scheme of Miller [40]. Different increments are assigned to different atom types. The estimates are accurate to within 1 to 3%. The method does not distinguish isomers of any type.

4.4.4. Molar Volume

Molar volume belongs to a group of "bulkiness" parameters whose dimensions are volume. The group includes molar refractivity as well as parachor. There is a significant correlation between the Van der Waals volume, V_w, according to Bondi [41] and molar refractivity as well as parachor whereas the correlation between V_w and molecular mass is relatively low [19]. Reference data for calculation of Van der Waals volume may be found elsewhere [41–43]. Modern computer packages employ the method described by Bodor *et al.* [44].

4.4.5. Van der Waals Surface Area

Modern computer graphics and calculation packages allow for visualization and estimation of Van der Waals surface area of complex molecules. In Fig. 4.2. exemplary surface area is given for acetazolamide as obtained by a commercially available software installed on a personal computer [45]. The grid calculation method applied is according to Bodor *et al.* [44] using the atomic radii of Gavezotti [46].

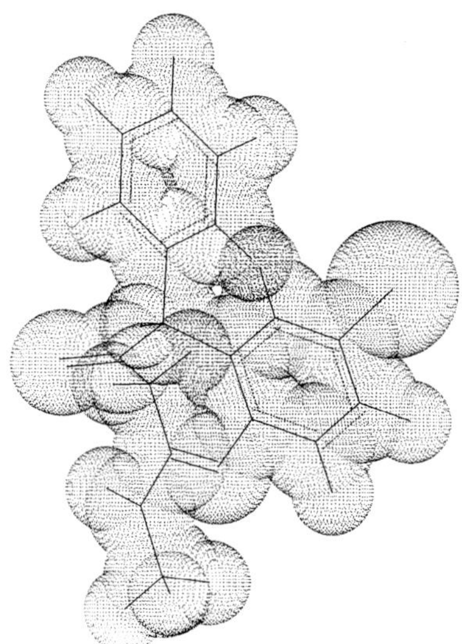

Figure 4.2. Van der Waals surface area for chlordiazepoxide.

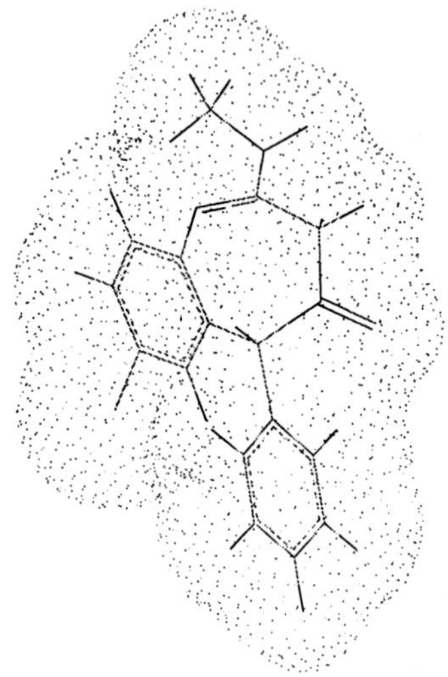

Figure 4.3. Water-accessible surface area for chlordiazepoxide.

4.4.6. Solvent-Accessible Surface Area

The parameter is obtained by rolling a spherical approximation of the solvent molecule over the Van der Waals surface of the studied molecule. When water is the solvent then it is normally approximated by a sphere of a radius of 14Å. The exemplary water-accessible surface area in Fig. 4.3 for chlordiazepoxide was obtained by a calculation program [45] according to the method of Still *et al.* [47].

4.4.7. Molecular Mechanics and Quantum Chemistry Indices

Structural descriptors supplied by theoretical chemistry are commonly used in QSRR analysis. At present, efficient computers and advanced molecular modeling software are standard equipment of every chemistry laboratory. Certainly, objections raised occasionally that the programs are used mechanically may be justified. However, in QSRR studies usually relatively large sets of solutes are considered. The calculations are normally performed at the same conditions for all the members of the set. It is true that the values of calculated parameters may deviate from the actual ones. Yet, even in such cases the QSRR derived need not be worthless if the error in calculation of these parameters for all the solutes is of a systematic nature. This is probably the reason why so many successful QSRR are reported in the literature involving molecular modeling indices as independent variables. Among the multitude of molecular modeling parameters employed in QSRR studies the most often used ones can be distinguished.

In 1984 Spanish researchers applied the total energy, E_T, in QSRR equations describing gas chromatographic retention of alkenes [48]. Analysing this and the following reports on relationships between retention data and E_T within homologous series Kaliszan [2] concluded that E_T reflected differences in the dispersive properties of solutes. In that respect E_T resembled carbon number, molar volume, refractivity and other "bulkiness"-related structural descriptors. The advantage of E_T over the other dispersion interactions parameters is that E_T differentiates individual isomers and even conformers. The E_T term has little or no real chemical meaning. It is likely that this total energy term is reflecting the nonspecific properties of the solutes indirectly and that other more chemically significant parameters are contained within this term.

The other quantum chemically derived well known structural parameter is the dipole moment. With this parameter was specifically concerned one of the first application of quantum mechanics in QSRR studies [49]. Although the QSRR [49] describing GC retention indices of phenol derivatives in terms of the calculated dipole moments were better than those employing the experimental dipole moments, the correlation of the observed and predicted retention parameters was not impressive. This led to an important observation of a generally poor performance of the total dipole moment as a retention descriptor. It was shown latter on [50] that submolecular polarity measures were more statistically significant parameters for describing retention. This observation is in agreement with early findings by Karger *et al.* [32] and Scott [51], who noted that molecules like 1,4-dioxane, which have a total dipole moment of zero, behave in chromatography as if composed of two polar subunits. Although the total dipole moment (actually its square) is still occasionally reported as a significant parameter in QSRR equations [52], it cannot be used to describe all the specific polar features of the compounds.

To find a more suitable for QSRR measure of chromatographic polarity of solutes Ośmiałowski *et al.* [53] considered electron excess charges on individual atoms in a molecule. In effect, a submolecular polarity parameter Δ has been proposed.

To determine Δ, electron densities on each atom in the molecule are considered and two atoms are found with highest electron excess and deficiency. The difference in electron densities for the two atoms gives Δ. In the two-parameter regression equation relating the Kováts index of a series of amines to E_T and Δ, the term Δ appeared significant at the $p < 0.006$ significance level.

Of similar nature to Δ may be considered the structural descriptors proposed by Buydens *et al.* [54]. In studies on the prediction of GC retention indices of mono- and bifunctional aliphatic compounds significant appeared the sum of the absolute values of the charges in a given molecule and the sum of the absolute values of the charges of the atoms constituting the functional group and the atoms in the α position to this functional group.

The quantum chemically calculated energies of the highest occupied molecular orbital (HOMO) and the lowest unoccupied molecular orbital (LUMO) called natural interest of QSRR students, especially if the electron pair donor-electron pair acceptor intermolecular interactions were expected to be decisive for retention. Even if the statistical significance of the orbital energy terms in the first reported QSRR [48,55] appears questionable latter publications (*e.g.*, ref. [53]) demonstrated usefullness of descriptors.

There are numerous molecular modeling derived structural descriptors used only occasionally in QSRR studies. For example Katritzky *et al.* [56] in their studies on prediction of GC retention times employed 24 quantum chemical descriptors of solutes. Within the group were: the maximum σ–π bond order, the maximum exchange energy for a C–H bond, the highest normal mode vibrational frequency, *etc.*

4.4.8. Structural Descriptors Derived from Molecular Formulae

4.4.8.1. Molecular Shape Parameters

In most standard analytical chromatographic systems the steric effects on retention are generally of minor importance in comparison to differences resulting from polar and dispersive interactions. This is obviously not the case with the affinity chromatography. In the affinity chromatography steric fitting is decisive for the process of molecular recognition. This steric fitting can rarely be described by a single quantity, however. Relevant QSRR will be discussed in Chapter 12.

Even in some classical chromatographic systems molecular shape effects of solutes manifest themselves clearly. For example, isomers of polycyclic aromatic hydrocarbons (PAH) can be separated by GC on nematic stationary phases. A good quantitative description of this separation was achieved after the introduction of the length-to-breadth ratio as a structural parameter. This shape parameter η is defined [57] as the ratio of the longer to the shorter side of a rectangle having a minimum area which can envelop a molecule (Fig. 4.4).

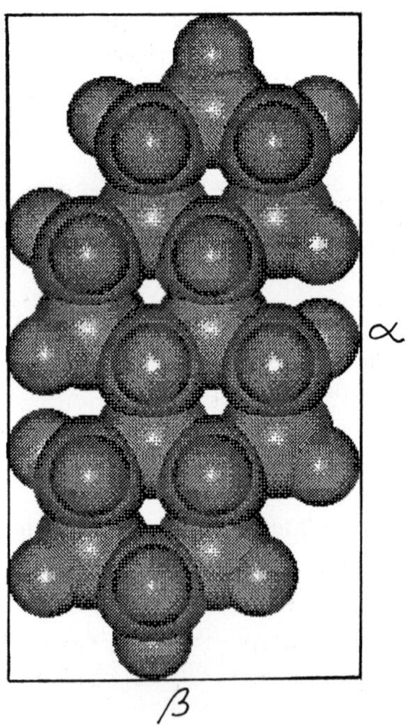

Figure 4.4. Determination of the shape parameter $\eta = \alpha/\beta$ for chrysene. By a and b are denoted the approximate lengths of carbon–carbon and carbon–hydrogen bonds (1.4 and 1.1 Å, respectively) and r is the Van der Waals radius of the hydrogen atom. (Reprinted with permission from Radecki, A., Lamparczyk, H. and Kaliszan, R., *Chromatographia* 12, 595, 1979. Copyright Friedr. Vieweg & Sohn.)

Among the other steric parameters derived from molecular geometry the most often employed (especially in structure-biological activity relationship studies) are Verloop's STERIMOL length and width parameters [58].

There are seemingly interesting shape parameters proposed *ad hoc* to derive predictive QSRR equations for specific set of chromatographic data. Unfortunately, attempts to apply such parameters for another sets of solutes and/or chromatographic parameters usually fail. Such probably are the shadow area descriptors proposed by Rohrbaugh and Jurs [59]. These descriptors are defined as the cross-sectional areas of the three-dimensional molecular models of the structures projected onto three mutually perpendicular planes. Because the descriptor values are highly dependent on the orientation of the molecular structures prior to the calculation, the authors [59] proposed a systematic technique for structure orientation. Standardized shadow areas are computed by dividing the cross-sectional areas obtained above by the areas of the largest rectangle which can enclose the molecules in relevant orientation. The standardized shadow areas on *xy* and *yz* planes appeared significant regressors in QSRR describing GC retention indices of polyhalogenated biphenyls [34].

4.4.8.2. Graph Theoretical Indices

Molecular graphs represent chemical molecules as composed of two entities: graph vertices representing atoms and graph edges representing bonds. The common feature of graph theory-based structural descriptors is that they are derived by counting graph edges, vertices or both. Rather simple arithmetic operations are carried out to get numerical indices. These indices are supposed to comprise information on properties of the molecules.

There are two main areas of application of structural descriptors based on molecular graphs. The first is the differentiation of chemical structures for purposes of chemical documentation. The second is the determination of mathematical properties of chemical structures, which could next be related to physicochemical (biological) properties of chemical compounds [60].

From the point of view of QSRR the uniqueness in the values of graph theoretical index for every individual molecular graph is not necessary as opposed to the criteria of good documentation. Nevertheless, the good index should take into account the size, symmetry, branching, rings, multiple bonds and heteroatom characteristics of a complex molecule [61].

First applications of graph theoretical invariants in studies of structure-property relationships were reported by Wiener [62] half a century ago. However, it was after Randić [63] proposed a molecular connectivity index for characterization of molecular branching that a sudden expansion of studies in the area started.

The index originally introduced by Randić [63] has been calculated by means of the equation

$$^{1}\chi = \sum_{s=1}^{t} (D_i D_j)_s^{-1/2} \quad i \neq j, \tag{4.11}$$

where s refers to an edge in a hydrogen-suppressed molecular graph; t is the total number of edges, D_i and D_j represent values attributed to the adjacent atoms i and j, the superscript 1 on χ denotes the so-called first-order connectivity index.

Due to the works by Kier and co-workers [64], a generalized connectivity index concept has been elaborated. According to that concept, index ${}^1\chi$ is the specific case obtained by considering the edge i–j of length $h = 1$. The summation, like that applied for generating ${}^1\chi$, may be extended over all possible paths of length h,

$$ {}^h\chi = \sum_{s=1}^{t} (D_i D_j \cdots D_{h+1})^{-1/2}, \tag{4.12} $$

where s refers to a single path of length h while t is the total number of paths of length h in a graph. Calculation of zero- (${}^0\chi$), first- (${}^1\chi$) and second-order (${}^2\chi$) molecular connectivity indices for 2,3,4-trimethylpentane is illustrated in Fig. 4.5.

From a formal mathematical point of view the graph theoretical approach should be limited to graphs composed of identical vertices and identical edges. In chemistry it would correspond to molecular graphs of aliphatic hydrocarbons (assuming that one takes no account of hydrogen atoms). Rational seems to be an extension of the approach to molecules composed of a constant fragment and a variable hydrocarbon part, *e.g.*, aliphatic alcohols, fatty acids, *etc.* Certainly, real chemical structures of interest in QSRR studies are more diverse and comprise various heteroatoms and bonds. Thus, in practice some modifications of graph theoretical indices are applied.

To account for the nature of the atoms and the unsaturation of bonds Hall and Kier [65] instead D_i (the valency of a vertex i) proposed to use the atom connectivity

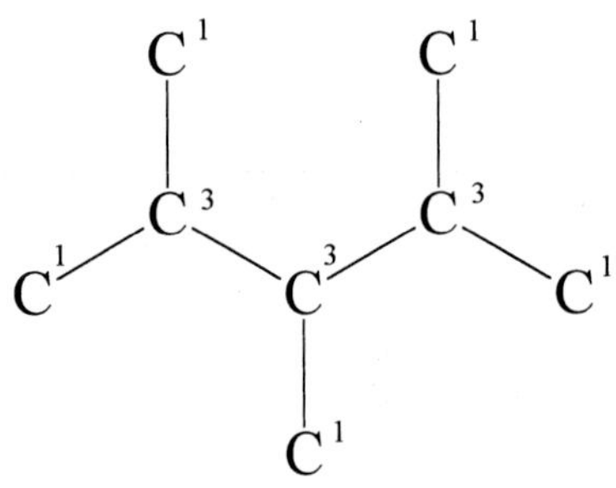

$$ {}^0\chi = 1^{-1/2} + 3^{-1/2} + 1^{-1/2} + 3^{-1/2} + 1^{-1/2} + 3^{-1/2} + 1^{-1/2} + 1^{-1/2} = 6.7320 $$

$$ {}^1\chi = (3 \times 3)^{-1/2} + (3 \times 1)^{-1/2} + (3 \times 3)^{-1/2} + (3 \times 1)^{-1/2} + (3 \times 3)^{-1/2} + \\ + (3 \times 1)^{-1/2} + (3 \times 1)^{-1/2} = 3.5534 $$

$$ {}^2\chi = (1 \times 3 \times 1)^{-1/2} + (1 \times 3 \times 3)^{-1/2} + (3 \times 3 \times 1)^{-1/2} + (3 \times 3 \times 3)^{-1/2} + \\ + (1 \times 3 \times 3)^{-1/2} + (3 \times 3 \times 1)^{-1/2} + (3 \times 3 \times 1)^{-1/2} + (1 \times 3 \times 3)^{-1/2} + \\ + (1 \times 3 \times 1)^{-1/2} = 3.3472 $$

Figure 4.5. Example calculation of molecular connectivity indices of different order.

Δ_i^v. This parameter is calculated by the formula

$$\Delta_i^v = Z_i^v - H_i, \tag{4.13}$$

where Z_i^v represents the number of valence electrons of atom i and H_i is the number of hydrogen atoms attached to this atom. Thus, the general formula for calculating the hth-order valence molecular connectivity is

$$^h\chi^v = \sum_{\text{paths}} (\Delta_i^v \Delta_j^v \cdots \Delta_{h+1}^v)^{-1/2}. \tag{4.14}$$

The heteroatom connectivities proposed by Kier and Hall [65,66] are given in Table 4.1. Example calculation of a valence molecular connectivity index is illustrated in Fig. 4.6.

There is a further differentiation of molecular connectivity indices based on the type of subgraph considered in the calculation. To the subgraphs of the types path (P), cluster (C), path-cluster (PC) and chain (CH) correspond the connectivity indices denoted χ_P, χ_{PC} and χ_{CH}, respectively.

Some recognition received also the average distance sum connectivity index J of Balaban calculated according to Barysz *et al.* [67] for molecules containing heteroatoms.

There are several reports on QSRR in which the so-called molecular shape indices or kappa indices of Kier [68] are employed. The kappa indices of first, second and

Table 4.1 Heteroatom connectivities Δ^v of Kier and Hall [66].

Group	Δ^v	*Group*	Δ^v
NH_4^+	1	–S–	0.944
NH_3	2		
$-NH_2$	3	=S= (with two single bonds, above and below)	3.58
–NH–	4	H_3O^+	3
=NH	4	H_2O	4
–N– (with single bond below)	5	–OH	5
≡N	5	–O–	6
=N– (pyridine)	5	=O	6
–N– (quaternary) (with single bonds above and below)	6	O (both nitro) O (both carboxylate)	6 6
–N= (nitro) (with ↓ below)	6	–F –Cl –Br –I	(–)20 0.690 0.254 0.085

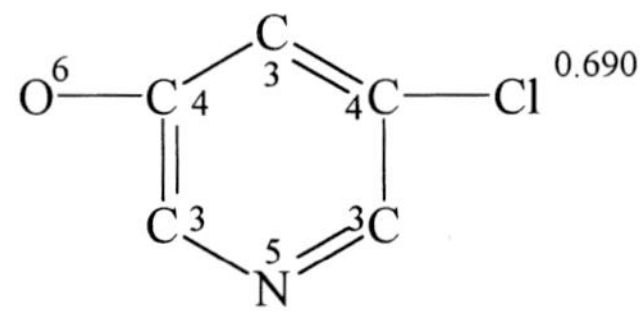

$$^1\chi^v = (3 \times 4)^{-1/2} + (4 \times 0.690)^{-1/2} + (4 \times 3)^{-1/2} + (3 \times 5)^{-1/2} + (5 \times 3)^{-1/2} + (3 \times 4)^{-1/2} + (4 \times 6)^{-1/2} + (4 \times 3)^{-1/2} = 2.4772$$

$$^2\chi^v = (3 \times 4 \times 0.690)^{-1/2} + (3 \times 4 \times 3)^{-1/2} + (4 \times 3 \times 5)^{-1/2} + (0.690 \times 4 \times 3)^{-1/2} + (3 \times 5 \times 3)^{-1/2} + (5 \times 3 \times 4)^{-1/2} + (3 \times 4 \times 3)^{-1/2} + (3 \times 4 \times 6)^{-1/2} + (4 \times 3 \times 4)^{-1/2} + (6 \times 4 \times 3)^{-1/2} = 1.8157$$

Figure 4.6. Example calculation of valence molecular connectivity.

third order, $^1\kappa$, $^2\kappa$ and $^3\kappa$ are calculated by the following formulae:

$$^1\kappa = \left(A + \sum\alpha\right)\left(A + \sum\alpha - 1\right)^2 \Big/ \left({}^1P + \sum\alpha\right)^2, \quad (4.15)$$

$$^2\kappa = \left(A + \sum\alpha - 1\right)\left(A + \sum\alpha - 2\right)^2 \Big/ \left({}^2P + \sum\alpha\right)^2, \quad (4.16)$$

$$^3\kappa = \left(A + \sum\alpha - 1\right)\left(A + \sum\alpha - 3\right)^2 \Big/ \left({}^3P + \sum\alpha\right)^2, \quad \text{when } A \text{ is odd}, \quad (4.17)$$

$$^3\kappa = \left(A + \sum\alpha - 3\right)\left(A + \sum\alpha - 2\right)^2 \Big/ \left({}^3P + \sum\alpha\right)^2, \quad \text{when } A \text{ is even}, \quad (4.18)$$

where ^nP is the number of paths for order of attribute n, A is the number of atoms and α is the increment of size contribution of a given atom relative to $C_{sp}3$. Possible paths of first, second and third order in graph of 2-phenybutane are indicated in Fig. 4.7.

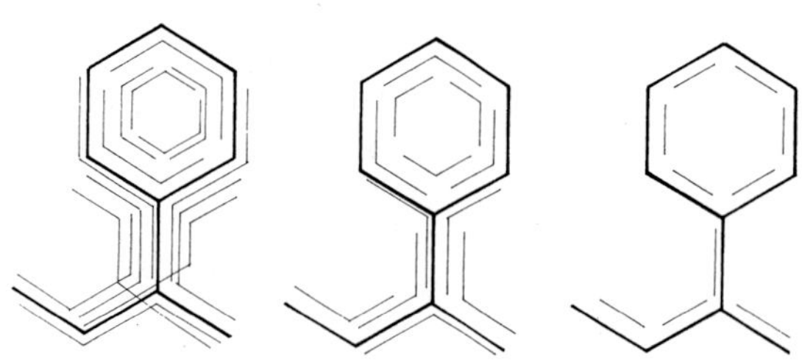

Figure 4.7. Paths of first (right), second (middle) and third (left) order in molecular skeleton of 2-phenyl-butane.

The most popular among the kappa indices is $^3\kappa$. According to Kier [68] $^3\kappa$ encodes structural features related to central branching.

Information content indices of neighbourhood symmetry, IC, of different order are also calculated from structural formulae [69]. These indices are calculated by using the general information theory equation (Shannon equation) from probabilities of finding equivalent atoms (or patterns of atoms) in a given structural formula:

$$\mathrm{IC}_n = -\sum_i (p_i \log_2 p_i), \tag{4.19}$$

where n is the order of IC index and p_i is the probability of finding equivalent atoms (EA) with regard to their neighbourhood in a structural formula. To calculate p_i values at first sets of digits are assigned to individual atoms (numbered with subscripts) which characterize their closest neighbours. For example, in Fig. 4.8 the set of digits $1^1 1^4 2^4$ is at the C_2 atom, which means that C_2 is connected by a single bond with a one-valence atom (H_1), by another single bond with a four-valence atom (C_3) and by a double bond with a four-valence atom (C_1). Of 16 atoms in the structural formula, seven are carbons (C_1–C_7). Five of these (C_2–C_6) have identical first-order neighbourhood. Thus, probability of finding these five atoms (C_2–C_6) of

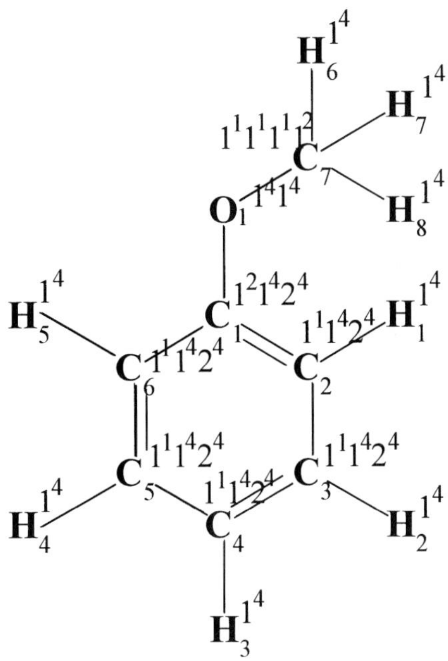

Figure 4.8. Sets of digits characterizing closest neighbours of atoms numbered from 1 to 16 in formula of anisole.

the same first-order neighbourhood is $p_1 = 5/16$. However, a second-order neighbourhood analogous to C_2 has only the atom C_6. These two atoms provide $p_2 = 2/16$ for calculating IC_2. In Table 4.2 probabilities p_2 are assigned to individual equivalent atoms (EA) regarding their zeroth-, first- and second-order neighbourhood.

There are several other molecular formula-derived structural descriptors reported incidentally in QSRR studies. They are to comprise some systematic information related to interatomic distances and to individual atoms connections. Generally, they do not utilize information concerning the electronic structure of the solutes. At present, when quantum chemical calculations and molecular modeling can be done routinely, the topological indices based on such obtained data may become readily accessible. Most probably, the quantum chemical data may be transformed into various topological indices of value for specific discriminative purposes. An example is the topological electronic index T^E calculated from the formula [70]

$$T^E = \sum_{i,j} (|q_i - q_j| / r_{i,j}^{\,2}), \quad i \neq j, \tag{4.20}$$

where q_i and q_j are electron excess charges on pair of atoms separated by the distance $r_{i,j}$ (in Å). The distances $r_{i,j}$ between each pair of atoms are calculated based on the atom coordinates in three-dimensional space as obtained by molecular modeling. For every pair of atoms the absolute value of the excess charge difference is divided by the square of the respective interatomic distance. The resulting numbers are summed for all possible atomic pairs in the molecule. The method of calculation is illustrated in Fig. 4.9.

The topological electronic index T^E has occasionally been reported in QSRR studies [70–72].

Molecular formula-derived solute descriptors are easy to calculate and thus appear attractive to QSRR students. Some of them certainly differentiate molecular formulae. However, in spite of numerous reports on their application in studies of quantitative structure-property relationships it remains an open question whether these descriptors unambiguously differentiate the properties of compounds represented by individual formulae.

Table 4.2 Probabilities p_0, p_1 and p_2 of finding of equivalent atoms, EA, of identical zeroth-, first- and second-order neighbourhood in Fig. 4.8. The respective information content indices IC_0, IC_1 and IC_2 were calculated by Eq. (4.19).

IC_0		IC_1		IC_2	
EA	p_0	*EA*	p_1	*EA*	p_2
$C_1 \ldots C_7$	7/16	C_1	1/16	C_1	1/16
$H_1 \ldots H_8$	8/16	$C_2 \ldots C_6$	5/16	C_2, C_6	2/16
O_1	1/16	C_7	1/16	$C_3 \ldots C_5$	3/16
		$H_1 \ldots H_8$	8/16	C_7	1/16
		O_1	1/16	$H_1 \ldots H_5$	5/16
				$H_6 \ldots H_8$	3/16
				O_1	1/16
$IC_0 = 1.2718$		$IC_1 = 1.7744$		$IC_2 = 2.5550$	

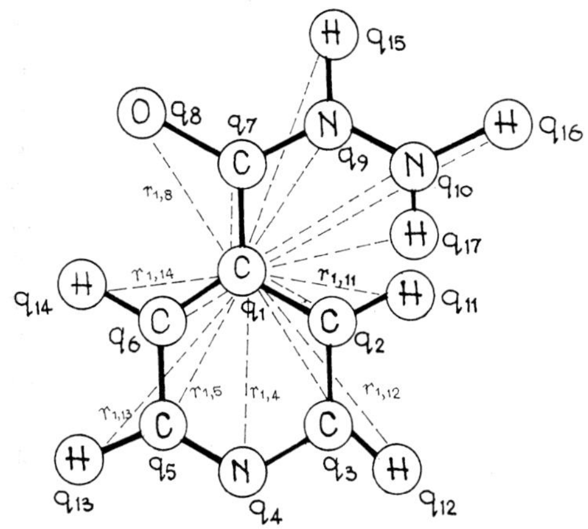

Figure 4.9. Electron excess charges q on individual atoms in isonicotin hydrazide. Interatomic distances between carbon C_4 and remaining atoms in the molecule are indicated by broken lines.

4.4.9. Empirical Physicochemical Parameters in QSRR Analysis

QSRR involving the experimental physicochemical parameters as independent variables are valuable from the point of view of elucidation of mechanism of retention operating in a specific chromatographic systems. The problem is, however, with the availability of sufficiently large for statistical analysis sets of comparable physicochemical data. In such a situation QSRR are rather used to evaluate physicochemical properties than *vice versa*.

The limited adequacy of the experimental dipole moments for QSRR analysis has been discussed in Section 4.4.7.

Dependence of GC retention parameters on the boiling point of congeneric solutes has been noted since the introduction of this method of analysis. An interesting QSRR study of this kind is by Bermejo *et al.* [73].

Applications of the $ET(30)$ solvatochromic solvent polarity parameter in QSRR have been discussed by Dorsey and Khaledi [74].

Empirical in nature are also the substituent and/or fragmental contributions to retention parameters. Numerous predictive QSRR based on this approach have been reported by Smith [75] and by Wells and Clark [76].

Solute acidity parameter pK_a is used in QSRR equations describing reversed-phase liquid chromatographic retention mostly as a correction factor accounting for dissociation of solutes.

The most often used in QSRR experimental physicochemical parameter is the logarithm of the *n*-octanol-water partition coefficient, $\log P$. *n*-Octanol-water partition coefficients are determined in "shake-flask" experiments. As good detailed description of such typical experiment can be found elsewhere [77]. Collections of $\log P$ data for tens of thousands of solutes are available [15,35,78–80].

There are numerous reports on QSRR in which $\log P$ is a single retention descriptor (mostly in reversed-phase HPLC and TLC). In the review published in 1993 Lambert [81] reports on at least 116 such QSRR aimed at chromatographic determination of solute hydrophobicity. A separate chapters in this book are devoted to the chromatographic hydrophobicity parameters and their applications in medicinal chemistry, molecular pharmacology and environmental analysis (Chapters 11 and 12).

Other than *n*-octanol-water partitioning data have also been occasionally used for description of chromatographic retention. An interesting example is by Carr *et al.* [82] who discuss relationship between the logarithm of chromatographic capacity factor determined by Tsukahara *et al.* [83] and the logarithm of partition coefficient of solutes between dodecane and the methanol-water solvent.

4.5. General Remarks on QSRR Descriptors

The above discussed structural descriptors are most often used in QSRR analysis. There are hundreds of others which use has been reported. From the purely statistical point of view these are accepted which provide formally significant QSRR. However, the number of descriptors which can be assigned to an individual solute is practically unlimited, especially if various transformations and combinations of particular descriptors are induced. As pointed out by Topliss and Edwards [84], if enough parameters are screened even random numbers show correlations. Thus, it is recommended that the descriptors used are interpretable in physical terms. One can believe, for example, that the third-order kappa index of Kier, ${}^{3}\kappa$, might encode the "general shape of the molecules". However arriving at a QSRR equation in which the term $({}^{3}\kappa)^{3}$ appears statistically significant one has to ask what the "general shape" raised to the third power means.

Of course, in chemistry one seldom deals with clearly isolated systems and the truly one-dimensional variables. Well designed structural descriptors and advanced chemometric methods of data analysis help to identify the chemical sense comprised in meaningful QSRR equations.

Having the above in mind one should start QSRR analysis with a set of solute descriptors which have rather well established chemical meaning. In Table 4.3 a classification of more commonly used in QSRR analysis structural descriptors is given [85,86].

Solute bulkiness descriptors may be ascribed to the ability of a compound to participate in non-specific, dispersive inter molecular interactions with the components of a chromatographic system. These interactions are very important for chromatographic separations and the bulkiness parameters are usually found significant in QSRR equations. Obviously, the bulkiness parameters are decisive in description of separation of closely congeneric solutes (*e.g.*, carbon number suffices to differentiate the members of homologous series). On the other hand, when dealing with a set of solutes of the same size (like isomers), they may appear insignificant in QSRR equations. This does not mean that dispersive interactions are meaningless for separation of congeners. They are just closely similar, if not identical, and the respective term in QSRR equation looses statistical significance.

Molecular polarity is much more difficult to quantify. The descriptors used are expected to account for differences among the solutes regarding their dipole–dipole, dipole–induced dipole, hydrogen bonding and EPD–EPA interactions. To find a good descriptor of specific interactions affecting a given chromatographic separation

Table 4.3 Basic groups of structural descriptors used in QSRR studies.

Molecular bulkiness-related non-specific descriptors	*Molecular polarity-related (electronic) descriptors*
Carbon number	Dipole moments
Molecular mass	Atomic and fragmental electron excess charges
Refractivity	Orbital energies of HOMO and LUMO
Polarizability	Partially charged areas
Van der Waals volume and area	Local dipoles
Solvent-accessible volume and area	Submolecular polarity parameters
Total energy	
Molecular geometry-related (shape) descriptors	*Molecular graph-derived (topological) parameters*
Length-to-breadth ratio	Connectivity indices
STERIMOL parameters	Kappa indices of skape
Moments of inertia	Information content indices
Shadow area parameters	Topological electronic index
Physicochemical empirical and semiempirical parameters	*Indicator variables*
Hammett constants	Zero-one indices
Hydrophobic constants	
Taft steric constants	
Hydrophobic fragmental parameters	
Solubility parameters	
Solvatochromic parameters	
Partition coefficients	
Boiling temperatures	

is normally not an easy task. The more so that changes in solute polarity cause also the changes in their ability to participate in dispersive interactions and affect solute geometry.

The geometry (shape) parameters are difficult to quantify one-dimensionally. Single numbers reflecting the shape differences are adequate only in case of rigid, planar solutes. They become significant in QSRR equations when a series of solutes considered comprises compounds of simiar size and polarity. The significance of individual shape parameters (usually designed *ad hoc*) is pronounced in QSRR analysis of retention data generated by affinity chromatography.

The physical meaning of molecular graph-derived descriptors is not ever clear a *priori*. It is rather that a good QSRR equation allows for assigning the physical meaning to the individual topological indices.

Empirical physicochemical parameters have a good informative value for determining the mechanism of retention operating in a given chromatographic system studied. For example, good correlations of retention with the hydrophobicity parameters are treated as proofs of partition chromatography mechanism. The real value of QSRR involving physicochemical (and biological) parameters is that the chromatographic systems yielding such QSRR may be used to conveniently evaluate these physicochemical (biological) parameters for any other chromatographed solutes.

Indicator values 0–1 assigned to solutes depending on the presence–absence of a given structural feature help sometimes to identify numerical descriptor of structure of physical significance. Unfortunately, often such indicator descriptors just help to get statistically (but not necessarily physically) acceptable QSRR equations.

The list of the most commonly used descriptors discussed above is not usually sufficient to derive the requested QSRR equation for the actual chromatographic data. There is a space for invention of individual researchers. As a matter of fact, the QSRR analysis is expected to identify the good structural descriptors suitable for other kinds of structure-property relationships.

References

1. Giddings, J.C., "*Unified Separation Science*", Wiley, New York, 1991, p. 24.
2. Kaliszan, R., "*Quantitative Structure-Chromatographic Retention Relationships*", Wiley, New York, 1987, p. 2.
3. Tomlinson, E., *British Pharmaceutical Conference Science Award Lecture*, Brighton, 1981.
4. Leffler, J.E. and Grunwald, E., "*Rates and Equilibria of Organic Reactions*", Wiley, New York, 1963.
5. Shorter, J., "*Correlation Analysis in Organic Chemistry: An Introduction to Linear Free-Energy Relationships*", Clarendon, Oxford, 1973, p. 64.
6. Melander, W., Campbell, D.E. and Horváth, C., *J. Chromatogr.* **158**, 213 (1978).
7. Martin, A.J.P., *Biochem. Soc. Symp.* **3**, 4 (1950).
8. Green, J., Marcinkiewicz, S. and McHale, D., *J. Chromatogr.* **10**, 158 (1963).
9. Tomlinson, E., Poppe, H. and Kraak, J.C., *Int. J. Pharm.* **7**, 225 (1981).
10. Hansch, C., Sammes, P.G. and Taylor, J.B. (Eds.), "*Comprehensive Medicinal Chemistry*", Vols. **1–6**, Pergamon Press, Oxford, 1990.
11. Van de Waterbeemd, H. and Testa, B., in: "*Advances in Drug Research*", Vol. **16**, Testa B. (Ed.), Academic Press, London, 1987, p. 85.
12. Mager, P.P., *Sci. Pharm.* **48**, 117 (1980).
13. Seydel, J.K. and Schaper, K.-J., "*Chemische Struktur und biologische Aktivität von Wirkstoffen*", Verlag Chemie, Weinheim, 1979.
14. Hansch, C. and Fujita, T., *J. Am. Chem. Soc.* **86**, 1616 (1964).
15. Hansch, C. and Leo, A., "*Substituent Constants for Correlation Analysis in Chemistry and Biology*", Wiley, New York, 1979.
16. Iwasa, J., Fujita, T. and Hansch, C., *J. Med. Chem.* **8**, 150 (1965).
17. Tomlinson, E., *J. Chromatogr.* **113**, 1 (1975).
18. Kaliszan, R., *Crit. Rev. Anal. Chem.* **16**, 323 (1986).
19. Charton, M. and Motoc, I. (Eds.), "*Steric Effects in Drug Design*", Akademie Verlag, Berlin, 1983.
20. Haken, J. K., Nguyen, A. and Wainwright, M.S., *J. Chromatogr.* **178**, 471 (1979).
21. Sadek, P.C., Carr, P.W., Doherty, R.M., Kamlet, M.J., Taft, R.W. and Abraham, M.H., *Anal. Chem.* **57**, 2971 (1985).
22. Carr, P.W., Doherty, R.M., Kamlet, M.J., Taft, R.W., Melander, W. and Horváth, C., *Anal. Chem.* **58**, 2674 (1986).
23. Snyder, L.R., Carr, P.W. and Rutan, S.C., *J. Chromatogr. A* **656**, 537 (1993).
24. Park, J.H., Dallas, A.J., Chau, P. and Carr, P.W., *J. Chromatogr. A* **677**, 1 (1994).
25. Kamlet, M.J., Abboud, J.L.M. and Taft, R.W., *Prog. Phys. Org. Chem.* **13**, 591 (1981).
26. Kamlet, M.J., Abboud, J.L.M., Abraham, M.H. and Taft, R.W., *J. Org. Chem.* **48**, 2877 (1983).
27. Dimroth, K., Reichardt, C., Siepmann, T. and Bohlmann, F., *Liebigs Ann. Chem.* **661**, 1 (1963).
28. Kamlet, M.J., Doherty, R.M., Abraham, M.H., Marcus, Y. and Taft, R.W., *J. Phys. Chem.* **92**, 5244 (1988).
29. Leahy, D.E., Morris, J.J., Taylor, P.J. and Wait, A.R., *J. Chem. Soc. Perkin Trans. 2* **1992**, 705 (1992).
30. Abraham, M.H., Grellier, P.L. and McGill, R.A., *J. Chem. Soc. Perkin Trans. 2* **1992**, 797 (1987).
31. Hoy, K.L., *J. Paint Technol.* **42**, 76 (1970).
32. Karger, B.L., Snyder, L.R. and Eon, C., *J. Chromatogr.* **125**, 71 (1976).
33. Tijssen, R., Billiet, A.H. and Schoenmakers, P.J., *J. Chromatogr.* **122**, 185 (1976).
34. Hasan, M.N. and Jurs, P.C., *Anal. Chem.* **62**, 2318 (1990).
35. Rekker, R.F., "*The Hydrophobic Fragmental Constant: Its Derivatization and Application*", Elsevier, Amsterdam, 1977.
36. Vogel, A.J., Cresswell, W.T. and Leicester, J., *J. Phys. Chem.* **58**, 174 (1954).
37. Martin, Y.C., "*Quantitative Drug Design: A Critical Introduction*", Marcel Dekker, New York, 1978.

38. Dunn, W.J., *Eur. J. Med. Chem.* **12**, 109 (1977).
39. Ghose, A.K. and Crippen, G.M., *J. Chem. Inf. Comput. Sci.* **27**, 21 (1987).
40. Miller, K.J., *J. Am. Chem. Soc.* **112**, 8533 (1990).
41. Bondi, A., *J. Phys. Chem.* **68**, 441 (1964).
42. Moriguchi, J. and Kanada, Y., *Chem. Pharm. Bull.* **25**, 926 (1977).
43. Yang, G., Lien, E.J. and Guo, Z., *Quant. Struct. Act. Relat.* **5**, 12 (1986).
44. Bodor, N., Gabanyi, Z. and Wong, C., *J. Am. Chem. Soc.* **111**, 3783 (1989).
45. "*ChemPlus: Extensions for HyperChem*", Hypercube, Inc., Canada, 1994.
46. Gavezotti, A., *J. Am. Chem. Soc.* **10**, 5220 (1983).
47. Still, W.C., Tempczyk, A., Hawley, R.C. and Hendrickson, T., *J. Am. Chem. Soc.* **112**, 6127 (1990).
48. Garcia-Raso, A., Saura-Calixto, F. and Raso, M.A., *J. Chromatogr.* **302**, 107 (1984).
49. Kaliszan, R. and Höltje, H.-D., *J. Chromatogr.* **234**, 303 (1982).
50. Kaliszan, R., Ośmiałowski, K., Tomellini, S.A., Hsu, S.-H., Fazio, S.D. and Hartwick, R.A., *Chromatographia* **20**, 705 (1985).
51. Scott, R.P.W., *J. Chromatogr.* **122**, 35 (1976).
52. Ong, V.S. and Hites, R.A., *Anal. Chem.* **63**, 2829 (1991).
53. Ośmiałowski, K., Halkiewicz, J., Radecki, A. and Kaliszan, R., *J. Chromatogr.* **346**, 53 (1985).
54. Buydens, L., Massart, D.L. and Geerlings, P., *Anal. Chem.* **55**, 738 (1983).
55. Nondek, L. and Ponec, R., *J. Chromatogr.* **294**, 175 (1984).
56. Katritzky, A.R., Ignatchenko, E.S., Barcock, R.A., Lobanov, V.S. and Karelson, M., *Anal. Chem.* **66**, 1799 (1994).
57. Kaliszan, R., Lamparczyk, H. and Radecki, A., *Biochem. Pharmacol.* **28**, 123 (1979).
58. Verloop, A., Hoogenstraaten, W. and Tipker, J., in: "*Drug Design*", Vol. 7, Ariëns, E.J. (Ed.), Academic Press, New York, 1976.
59. Rohrbaugh, R.H. and Jurs, P.C., *Anal. Chem. Acta* **199**, 99 (1987).
60. Ośmiałowski, K. and Kaliszan, R., *Quant. Struct.-Act. Relat.* **10**, 125 (1991).
61. Bertz, S.H., *J. Am. Chem. Soc.* **103**, 3599 (1981).
62. Wiener, H., *J. Phys. Chem.* **52**, 1082 (1948).
63. Randić, M., *J. Am. Chem. Soc.* **97**, 6609 (1976).
64. Kier, L.B. and Hall, L.H., "*Molecular Connectivity in Chemistry and Drug Research*", Academic Press, New York, 1976.
65. Hall, L.H. and Kier, L.B., *J. Pharm. Sci.* **66**, 642 (1977).
66. Kier, L.B. and Hall, L.H., "*Molecular Connectivity in Structure-Activity Analysis*", Research Studies Press, Letchworth, 1986.
67. Barysz, M., Jashari, G., Lall, R.S., Srivatstave, V.K. and Trinajstić, N., in: "*Chemical Applications of Topology and Graph Theory*", King, R.B. (Ed.), Elsevier, Amsterdam, 1983.
68. Kier, L.B., *Quant. Struct-Act. Relat.* **5**, 1 (1986).
69. Sarkar, R., Roy, A.B. and Sarkar, P.K., *Math. Biosci.* **39**, 299 (1978).
70. Ośmiałowski, K., Halkiewicz, J. and Kaliszan, R., *J. Chromatogr.* **361**, 63 (1986).
71. Hansson, G. and Ahnoff, M., *Chemom. Intell. Lab. Sys.* **17**, 223 (1992).
72. Hansson, G. and Ahnoff, M., *J. Chromatogr. A* **666**, 505 (1994).
73. Bermejo, J., Blanco, C.G. and Guillén, M.D., *J. Chromatogr.* **351**, 425 (1986).
74. Dorsey, J.G. and Khaledi, M.G., *J. Chromatogr. A* **656**, 485 (1993).
75. Smith, R.M., *J. Chromatogr. A* **656**, 381 (1993).
76. Wells, M.J.M. and Clark, C.R., *Anal. Chem.* **64**, 1660 (1992).
77. De Voogt, P., Wegner, J.W.M., Klamer, J.C., Van Zijl, G.A. and Govers, H., *Biomed. Environ. Sci.* **1**, 194 (1988).
78. Craig, P.N., in: "*Comprehensive Medicinal Chemistry*", Vol. 6, Hansch, C., Sammes, P.G. and Taylor, J.B. (Eds.), Pergamon Press, Oxford, 1990.
79. Hansch, C. and Leo, A., "*Pomona College Medicinal Chemistry Project* log *P Data Bank*", Pomona College, 1983.
80. Van de Waterbeemd, H., in: "*Hydrophobicity of Organic Compounds, Booksoft Vol. 1*", Darvas, F. (Ed.), Compudrug, Budapest, 1986.
81. Lambert, W.J., *J. Chromatogr. A* **656**, 469 (1993).
82. Carr, P.W., Li, J., Dallas, A.J., Eikens, D.I. and Tan, L.C. *J. Chromatogr. A* **656**, 113 (1993).
83. Tsukahara, S., Saitoh, K. and Suzuki, N., *Anal. Sci.* **9**, 71 (1993).
84. Topliss, J.G. and Edwards, R.P., *J. Med. Chem.* **22**, 238 (1979).
85. Kaliszan, R., *Anal. Chem.* **64**, 19A (1992).
86. Kaliszan, R., *J. Chromatogr. A* **656**, 417 (1993).

5 Chromatographic Retention Parameters as Dependent Variables in QSRR Studies

5.1. General Characteristics

Quantitative structure-retention relationships (QSRR) are assumed to result from the general extrathermodynamic concept of the linear free-energy relationships (LFER). Theoretical basis to this assumption are discussed in the first two chapters of this book.

Chromatographic data normally used in QSRR studies may have some objections regarding their thermodynamic analysis. However, as Prausnitz [1] observed it: "Classical thermodynamics is revered, honoured and admired, but in practice it is inadequate".

Chemical compounds are normally obtained for their requested properties and not to determine the thermodynamics of chemical systems. From the point of view of a search for better drugs, pesticides, detergents, building materials, *etc*., thermodynamics appears to be a knowledge acquired after the fact. Of course, it does not mean that thermodynamics is useless in chromatography. It forms the basis of separation, it provides rational approaches to increase the performance of separation systems, it guides the efforts to optimize separations. No chromatographic research or creative application is possible without a solid thermodynamic background.

When preparing or selecting chromatographic data for QSRR analysis one should have the basic thermodynamic rules in mind. Although QSRR are derived statistically and the structural descriptors used are often of obscure thermodynamic significance, the more reliable the chromatographic dependent variables are the better is the chance to arrive at physically meaningful relationships. The basic chemometrics idea, to extract maximum information from the analysis of chemical data sets, can be realized if the analysed data are reliable, precise and reproducible.

Wide interest of chemometricians and chromatographers in QSRR is mostly due to the fact that chromatography can readily yield a great amount of comparable property data for large series of diversified structures. In a chromatographic process all conditions may be kept constant. Thus, solute structure becomes the single (however multidimensional) independent variable in the system. QSRR studies appear to be the best method for testing the applicability of individual structural parameters for property description. The knowledge and skill gained from QSRR analysis may be applicable to other structure-property relationship studies, including the quantitative structure-biological activity relationship (QSAR) studies [2].

5.2. Gas Chromatographic Retention Indices

The most commonly used index in gas chromatographic QSRR is the Kováts retention index. To uniformly report GC retention data Kováts [3] proposed a system using *n*-alkanes as a series of standards. Later on, following the suggestion by Ettre [4] that the originally proposed retention volumes could be replaced by adjusted retention times, Kováts [5] introduced his known formula

$$I_i(T) = 100(\log t'_{Ri} - \log t'_{Rz})/(\log t'_{R(z+1)} - \log t'_{Rz}) + 100z, \tag{5.1}$$

where $I_i(T)$ is the Kováts retention index characteristic of a substance i chromatographed on a given column (stationary phase) at a definite temperature, t'_{Rz} is the adjusted retention time of a homologue standard with carbon number z, $t'_{R(z+1)}$ is an analogous parameter for another standard with carbon number $(z+1)$ and t'_{Ri} is the adjusted retention time of substance i.

The adjusted retention time t'_R which is independent of the equipment used is related to the total retention time directly measured in GC, t'_R, and to the dead time, t'_M, which is dependent on the system flow rate as well as the void volume of the apparatus, by the following equation:

$$t_R = t_M + t'_R. \tag{5.2}$$

Actually, the adjusted retention time, $t'_{R'}$ is the quantity that could be related to solute structure. However, although independent of the equipment used, the adjusted retention time is still dependent on several variables such as column temperature, flow rate, pressure drop and stationary phase. When adjusted retention times are used to calculate Kováts indices, the retention parameters are obtained which depend only on the column temperature and stationary phase used [5].

To calculate the adjusted retention time from the measured retention time by Eq. (5.2) the dead time must be known for a given column. Various methods of determining dead time are discussed in literature [6–8].

Kováts indices are highly reproducible. With a well designed experimental technique and an accurate timing mechanism, an interlaboratory reproducibility of one unit for larger values of Kováts indices and two units for indices below 400 is possible [8,9].

To attain conformability of various Kováts indices the temperature dependence has to be taken in mind. A temperature dependence of less than one unit per degree applies in most instances [10]. The dependence of retention indices on column temperature can be determined by the method described by Takács *et al.* [11].

Instead of Kováts indices sometimes in QSRR studies are used the logarithms of retention volumes of solutes.

GC retention indices are probably the best suitable for QSRR chromatographic retention parameters. Even if the actually considered data are not free of errors, these errors appear of secondary importance for QSRR, especially when one has in mind a generally low precision of the structural descriptors data.

5.3. Thin-Layer Chromatographic Parameters

Thin-layer chromatographic (TLC) retention parameters have been the first to be used in what is now called QSRR analysis. Classical TLC data were of rather limited reproducibility. The use of well-defined small-diameter particles of stationary phase materials and a better knowledge of the parameters which determine the efficiency of chromatographic systems have given rise to the high-performance TLC (HPTLC), by analogy with HPLC. Although TLC is not capable of performances comparable to those of column chromatography [12] it has several advantages from the point of view of QSRR studies. First, it requires considerably less laboratory equipment than HPLC. Second, tens of solutes can be simultaneously analysed on the same plate. Modern technical modifications of HPTLC, like the Soczewiński sandwich chamber [13], provide excellent control of separation conditions and drastically limit the use of expensive and environmentally polluting solvents. There is a much wider variability of eluents applied in TLC as compared to HPLC. Also, TLC offers the largest number of methods of detection [14].

Commercially available HPTLC plates are precoated with silica, chemically bonded silica stationary phases, cellulose and other materials. In case of silica materials, if low concentrations of a polar solvent are employed in the mobile phase, a significant proportion of the solvent is adsorbed by the stationary phase in the deactivation process. Thus, the concentration of polar solvent in the eluent front is significantly reduced. This produces a quasi-gradient elution effect [15]. Basically, this should only occur for low concentrations of polar solvent. The quasi-gradient elution is only significant for fast-moving spots which follow closely the solvent front for a significant proportion of the development.

The silica-coated TLC plates are generally not suitable for the analysis of very polar solutes. For such an analysis usually complex mixtures of solvents are required which are difficult to reproduce in practice [12]. Besides, the development of TLC plates by such mixtures gives rise to the phenomenon of demixing. The chromatograms of polar compounds then exhibit strong tailing and irregular spot shapes. This, in turn, causes problems in measuring migration distances.

When using HPTLC plates precoated with nonpolar chemically bonded stationary phases one encounters problems with wettability of the sorbent by eluents comprising higher proportions of water (buffer). Again there may be a selective adsorption of mobile phase components and a quasi-gradient elution for less retained solutes. Also buffer components may be selectively adsorbed. In contrast to HPLC, it is difficult to control the pH and ionic strength during the separation process in TLC. Experimental conditions allowing for precise estimation of RP TLC retention parameters are characteristic [16] by low organic modifier (ethanol) content in the mobile phase.

The retention parameter from TLC (and paper chromatography as well) which is normally used in QSRR is the R_M value defined by Bate-Smith and Westall [17]

$$R_M = \log(1/R_f - 1), \tag{5.3}$$

where R_f is the ratio of a distance passed by the solute to that attained by the solvent front. The R_M value is considered to be a LFER parameter.

Best accuracy in TLC is attained if R_f values are higher than 0.2 and lower than 0.8. This gives a range of R_M of less than one and a half decades, as compared to 3–4 decades in the case of HPLC [18]. In such a situation, it is often necessary to change the composition of the mobile phase for an individual solute in order to obtain reliable R_M values when one deals with a series of compounds of various retentive properties. This problem is overcome usually by determining R_M values at several compositions of binary eluents and linearly extrapolating (interpolating) the relationship between R_M and volume percent of one of eluent components to a fixed value. In the case of reversed-phase TLC extrapolation is usually performed to pure water (buffer) as a hypothetical eluent. This means zero percent of organic modifier used in individual TLC measurements. The such extrapolated R_M is usually denoted by R_M^0 or R_M^W.

The extrapolation to R_M^0 is based on the Soczewiński-Wachtmeister linear relationship [19] which, however, holds usually only at a more or less limited eluent concentration range. One would expect the R_M^0 value to be identical on a given stationary phase for a given solute, independent of the water-organic system used for R_M determinations and the subsequent extrapolation to R_M^0. However, this is seldom the case in practice. Still, analogously to HPLC, the extrapolated data are most often used in QSRR analysis of TLC data. Even if the extrapolated parameter was of a questionable physical meaning the extrapolation appears a convenient method of normalization of liquid chromatographic retention parameters. Since the first application in structure-property relationship studies by Boyce and Milborrow [20], followed by numerous reports by Biagi and co-workers [21,22] have the R_M^0 data been commonly used as quantitative TLC retention descriptors.

5.4. High-Performance Liquid Chromatographic Retention Parameters

QSRR involving HPLC retention parameters are at present the most extensively studied structure-chromatographic property relationships. There are numerous recent reviews demonstrating the value of such QSRR for elucidation of molecular mechanism of retention in diverse HPLC systems and for establishing of the chromatographic methods of convenient determinations of physicochemical and biological properties of solutes [23–28].

The LFER based retention parameter in HPLC is the logarithm of the phase capacity ratio or capacity factor k'. The capacity factor is defined as follows:

$$k' = (t_R - t_M)/t_M = (V_R - V_M)/V_M, \tag{5.4}$$

where t_R and V_R are the retention time and the retention volume, respectively, of the chromatographed solute. The symbols t_M and V_M denote the dead time and the dead volume of the column used.

The determination of the dead volume in HPLC presents both theoretical and practical problems. From the thermodynamic point of view, under the term dead volume one understands the total volume of all eluent components within the column bed. That is, not only the eluent in the interparticle space but also an additional volume of eluent within the pores of the particles of column packing [29]. However, any bonded stationary phase will preferentially adsorb certain components from eluents. The question arises whether these adsorbed eluent components are to be considered as part of the mobile or the stationary phase.

An unambiguous definition of what constitutes mobile phase and what constitutes stationary phase in the context of HPLC using porous packing materials is impossible. Nevertheless, if one is to make any thermodynamically valid measurements, which include simple measurements of column capacity ratios, one must provide a clear-cut definition of dead volume so that Eq. (5.4) for the capacity ratio can be used and data compared for various solutes. It should be stressed that the value used for V_M and its definition should ideally be the same for a single column for all eluents and should be readily determined with adequate precision. Unfortunately, the popular recipes to determine dead volume do not provide correct V_M data, except by chance [29].

The following methods of determination of V_M used in calculation of k' may be found in the literature: (i) V_M is the elution volume of a solvent disturbance or system peak obtained by injecting an eluent component; (ii) V_M is the elution volume of a non-ionized solute which gives the lowest retention volume and is small enough not to be sterically excluded; (iii) V_M is the elution volume of an isotopically labeled component of eluent, *e.g.*, ^2H_2O in the case of a reversed-phase packing material; (iv) V_M is the elution volume of salt or ion, usually a UV-absorbing ion; (v) V_M is the volume of liquid which the column contains (obtained, *e.g.*, by weighing the column full of liquid and then empty) less the volume of any adsorbed eluent components; (vi) V_M is the volume which, when substrated from elution volumes, V_{Rn} of a series of homologues, provides a linear dependence of $\log(V_{Rn} - V_M)$ against n, the number of carbon atoms in the homologues [6]. The present author recommends determining of V_M by flushing the column with any one-component eluent and measuring the elution volume of the isotopically labeled eluent sample. Deuterated eluent can be detected due to its different refraction coefficient. The method is not ideal, however, because of the known fact that the deuterium substitution may have an isotope effect on the distribution constants [30,31]. Most HPLC capacity factors used in QSRR analysis have, however, been calculated based on elution volumes of inorganic salts.

If the whole series of solutes which is subjected to QSRR analysis may clearly be separated at fixed isocratic conditions then $\log k'$ values thus obtained appear suitable for the study. One has to note, however, that the QSRR derived for given isocratic conditions need not to hold for data obtained at other isocratic conditions. This is because changes in proportion of component of the eluent often result in significant changes in relative retention or selectivity within a given series of solutes.

HPLC retention data for QSRR analysis are usually obtained by measuring $\log k'$ at several isocratic conditions and then by extrapolating (interpolating) the dependence of $\log k'$ on eluent composition to a common real or hypothetical mobile phase composition. Most QSRR reports deal with the reversed-phase HPLC employing binary water (buffer)-organic solvent systems. It is quite commonly assumed that the logarithm of capacity factor extrapolated to pure water (buffer) eluent, $\log k'_w$, is a special value and this parameter is normally used in QSRR analysis of RP HPLC data. Using extrapolated data one has to be cognizant theoretical background and limitations of such an approach.

Linear relationship of retention parameters on volume fractions of water in a binary aqueous eluent was analysed for planar-bed chromatography by Soczewiński and co-workers [19,32] and for HPLC by Snyder *et al.* [33]. It appeared soon that this dependence is nonlinear over the full composition range of the mobile phase. In any liquid-solid chromatographic system utilizing silica gel as the stationary phase and a mobile phase consisting of a mixture of a nonpolar and a polar solvent, Maggs [34] and Scott and Lawrence [35] observed a non-linearity of $\log k'$ against the volume fraction of the polar solvent at small concentration of the polar solvent.

The problem of quantitative relationships between retention parameters and mobile phase composition calls for much theoretical and practical interest. Interested readers are referred to a review by Valko *et al.* [36]. The approaches reported in QSRR publications are briefly discussed below.

The classical Soczewiński–Snyder model is represented by the following equation:

$$\log k' = \log k'_w - S\phi, \tag{5.5}$$

where k'_w corresponds to k' in pure water as mobile phase, S is constant for a given solute and a given RP HPLC system and ϕ is the volume fraction of the organic modifier. According to Eq. (5.5) zero percent of organic modifier corresponds to $\log k' = \log k'_w$.

The curvature often observed on plots of $\log k'$ versus ϕ led Schoenmakers *et al.* [37] to a quadratic relationship based on the solubility parameter theory

$$\ln k' = A\phi^2 + B\phi + C, \tag{5.6}$$

where A, B and C are constants for a given solute and a given chromatographic system.

Naturally, Eq. (5.6) provides better fit to experimental data than Eq. (5.5). The question is whether, with the practically measured k' data, the improvement is statistically significant or is just due to the presence of one additional term in Eq. (5.6) [38,39].

Some authors argue that a better linearization of dependence of retention parameters on eluent composition may be obtained after mathematical transformation of the variables. Thus, $\log k'$ *versus* $\log \phi$ [40] or $\log k'$ *versus* $\log(1/\phi)$ [41] relationships are occasionally employed. Recently, Lanin and Nikitin [42] derived relationship which rationalizes the plotting of the reciprocal of capacity factor, $1/k'$, against the mole fraction of the organic modifier.

According to some researchers the volume fraction of organic solvent is not the most appropriate parameter for regressing $\log k'$ data. Among the quantities proposed instead the widest interest called the actually measured solvent polarity [26,38,43]. To measure the polarity of organic-water solutions, Reichardt's dye ET (30) has been used due to its large solvatochromic shift. In view of subsequent reports [36,44,45] the improvement of the linearity of plots of $\log k'$ *versus* mobile phase composition is not as obvious when using ET (30) parameter instead of volume percent of organic modifier.

There are other more sophisticated approaches to the problem of dependence of HPLC retention on mobile phase composition (*e.g.*, refs. [46,47]). Application in QSRR studies found a comprehensive model of Jandera and co-workers [48,49].

In spite of many efforts, the relationships between retention and mobile phase composition are only approximate. One of the reason for this may be using of inappropriate dead volume values when calculating k' for different eluent compositions (see discussion regarding determination of dead volume above). The fact is that often the values of the logarithm of capacity factor extrapolated to a pure water eluent (the intercepts in Eq. (5.5)) are different from those determined experimentally (when it is possible). Besides, the $\log k'_w$ data are usually different when derived from aqueous systems modified with different organic solvents. This disagrees with Eq. (5.5). Because of this observation, some authors are inclined to believe that the extrapolation of capacity factors to 0% organic modifier is a manipulation and the value of $\log k'_w$ itself has no physical meaning [50,51].

Interpretation of $\log k'_w$ as the logarithm of HPLC capacity factor corresponding to a pure water (buffer) eluent is certainly misleading. The parameter is not devoid of merits as it may be regarded [52,53] as a means of normalizing retention data for a series of solutes tested. This feature is especially important from the point of view of QSRR studies.

5.5. Capillary Electrophoresis Retention Parameters

Electrophoretic mobility, μ, of spherical particles is described by a simple equation [54]

$$\mu = (z\Phi)/(6\pi\eta a N), \tag{5.7}$$

where z is the effective charge, Φ is charge per mole of protons, η is viscosity of the medium, a is the radius of the charged species and N is the Avogadro number.

A parameter normally measured in capillary electrophoresis (CE) is migration (retention) time, t. In a given CE system this parameter is inversely proportional to the electrophoretic mobility, μ. The μ (cm^3/Vs) is a normalized parameter, allowing for comparison of data obtained in different CE systems. If series of solutes is analysed under same conditions then $1/t$ and μ are equivalent.

Taking logarithms of both sides of Eq. (5.7) one gets

$$\log \mu = \log z - \log a + \text{constant}. \tag{5.8}$$

Thus, for a set of solutes analysed at the same conditions one should get regression equation describing $\log \mu$ with the coefficients close to 1 at the variables $\log z$ and $\log a$. Yet, such QSRR have not been found. As a matter of fact, in the simplest possible case of metal cations, for which the values of z and a are well defined, significant QSRR were derived with positive coefficients at z term and negative coefficients at the a term [55]. As regards organic solutes there are only semi-quantitative relationships observed for homologous or closely related solutes. Following Offord [56] several authors [57–60] found molecular mass raised to a power of $-2/3$ as a rough measure of the "effective surface area" of the migrating molecules. Another authors [61,62] related CE mobility of peptides to the number of amino acids in the polypeptide chain raised to a power 0.43. Also other reported attempts to derive QSRR for series of drugs cannot be considered successful [55,63].

Lack of success of QSRR analysis of CE data is disappointing but challenging. It may suggest unsuitability of the routinely determined mobility parameters as LFER descriptors of solute behaviour. As pointed out by Riekkola and co-workers [64,65] the repeatability of analyte migration times in CE is poor due mainly to non-repeatable electro osmotic flow velocity. The authors [64,65] developed a marker technique for high-accuracy identification by CE. The technique employs 2–4 markers of known electrophoretic mobilities which are used to determine the effective field strength and electro osmotic flow velocity of a system. The information about the last two variables, together with the electrophoretic mobilities of marker compounds, is used to determine the electrophoretic mobilities of unknown compounds.

References

1. Prausnitz, J.M., *Science* **205**, 759 (1979).
2. Kaliszan, R., *Anal. Chem.* **64**, 619A (1992).
3. Kováts, E., *Helv. Chim. Acta* **41**, 1915 (1958).
4. Ettre, L.S., *Anal. Chem.* **36**, 31A (1964).
5. Kováts, E., *Chimia* **22**, 459 (1968).
6. Kaliszan, R., "*Quantitative Structure-Chromatographic Retention Relationships*", Wiley, New York, 1987.
7. Ettre, L.S., *Chromatographia* **13**, 73 (1980).
8. Smith, R.J., Haken, J.K. and Wainwright, M.S., *J. Chromatogr.* **334**, 95 (1985).
9. Smith, R.J., Haken, J.K., Wainwright, M.S. and Madden, B.G., *J. Chromatogr.* **328**, 11 (1985).
10. Wehrli, A. and Kováts, E., *Helv. Chim. Acta* **42**, 2709 (1959).
11. Takács, J., Rockenbauer, M. and Olácsi, J., *J. Chromatogr.* **42**, 19 (1969).
12. Siouffi, A.M., Wawrzynowicz, T., Bressolle, E. and Guiochon, G., *J. Chromatogr.* **186**, 563 (1979).
13. Soczewiński, E., *J. Chromatogr.* **138**, 443 (1977).
14. Stahl, E., in: "*Chromatography, the State of the Art*", Kalasz, H. and Ettre, L.S. (Eds.), Akademiai Kiado, Budapest, 1985.
15. Scott, R.P.W., *J. Chromatogr.* **122**, 35 (1976).
16. Dross, K., Sonntag, C. and Mannhold, R., *J. Chromatogr.* **639**, 287 (1993).
17. Bate-Smith, E.C. and Westall, R.G., *Biochem. Biophys. Acta* **4**, 427 (1950).
18. Mirrlees, M.S., Moulton, S.J., Murphy, C.T. and Taylor, P.J., *J. Med. Chem.* **19**, 615 (1976).
19. Soczewiński, E. and Wachtmeister, C.A., *J. Chromatogr.* **7**, 311 (1962).
20. Boyce, C.B.C. and Milborrow, B.V., *Nature* **208**, 537 (1965).
21. Biagi, G.L., Barbaro, A.M., Gamba, M.F. and Guerra, M.C., *J. Chromatogr.* **41**, 371 (1969).
22. Pietrogrande, M.C., Bighi, C., Borea, P.A., Barbaro, A.M., Guerra, M.C. and Biagi, G.L., *J. Liq. Chromatogr.* **8**, 1711 (1985).
23. Kaliszan, R., *J. Chromatogr. A* **656**, 417 (1993).

24. Tan, L.C. and Carr, P.W., *J. Chromatogr. A* **656**, 521 (1993).
25. Dorsey, J.G. and Khaledi, M.G., *J. Chromatogr. A* **656**, 485 (1993).
26. Lambert, W.J., *J. Chromatogr. A* **656**, 469 (1993).
27. Sander, L.C. and Wise, S.A., *J. Chromatogr. A* **656**, 335 (1993).
28. Tanaka, N., Kimata, K., Hosoya, K., Miyanishi, H. and Araki, T., *J. Chromatogr. A* **656**, 265 (1993).
29. Knox, J.H. and Kaliszan, R., *J. Chromatogr.* **349**, 211 (1985).
30. Tanaka, N. and Thornton, E.R., *J. Am. Chem. Soc.* **99**, 7300 (1977).
31. Poppe, H., *J. Chromatogr. A* **656**, 19 (1993).
32. Soczewiński, E. and Matysik, G., *J. Chromatogr.* **32**, 458 (1968).
33. Snyder, L.R., Dolan, J.W. and Gant, J.R., *J. Chromatogr.* **165**, 3 (1979).
34. Maggs, R.E., *J. Chromatogr. Sci.* **7**, 145 (1969).
35. Scott, R.P.W. and Lawrence, J.G., *J. Chromatogr. Sci.* **8**, 619 (1970).
36. Valko, K., Snyder, L.R. and Glajch, J.L., *J. Chromatogr. A* **656**, 501 (1993).
37. Schoenmakers, P.J., Billiet, H.A.H., Tijssen, R. and De Galan, L., *J. Chromatogr.* **149**, 519 (1978).
38. Johnson, B.P., Khaledi, M.G. and Dorsey, J.G., *Anal. Chem.* **58**, 2354 (1986).
39. Sadlej-Sosnowska, N. and Śledzińska, J., *J. Chromatogr.* **595**, 53 (1992).
40. Taylor, R.B., Ochekpe, N.A. and Wangboonskul, J., *J. Liq. Chromatogr.* **12**, 1645 (1989).
41. Kaibara, A., Hirose, M. and Nakagawa, T., *Chem. Pharm. Bull.* **39**, 720 (1991).
42. Lanin, S.N. and Nikitin, Y.S., *J. Chromatogr.* **537**, 33 (1991).
43. Sadek, P.C., Carr, P.W., Doherty, R.M., Kamlet, M.J., Taft, R.W. and Abraham, M.H., *Anal. Chem.* **57**, 2971 (1985).
44. Minick, D.J., Frenz, J.H., Patrick, M.A. and Brent, D.A., *J. Med. Chem.* **31**, 1923 (1988).
45. Michels, J.J. and Dorsey, J.D., *J. Chromatogr.* **499**, 435 (1990).
46. Melander, W.R., Chen, B.-K. and Horváth, C., *J. Chromatogr.* **318**, 1 (1985).
47. Geng, X. and Regnier, F.E., *J. Chromatogr.* **332**, 147 (1985).
48. Jandera, P., Colin, H. and Guiochon, G., *Chromatographia* **16**, 132 (1982).
49. Jandera, P., *J. Chromatogr. A* **656**, 437 (1993).
50. Miyake, K., Mizuna, N. and Terada, H., *J. Chromatogr.* **439**, 227 (1988).
51. Sherblom, P.M. and Eganhouse, R.P., *J. Chromatogr.* **454**, 37 (1988).
52. Braumann, T. and Jastorff, B., *J. Chromatogr.* **350**, 105 (1985).
53. Braumann, T., *J. Chromatogr.* **373**, 191 (1986).
54. Li, S.F.Y., "*Capillary Electrophoresis. Principles, Practice and Applications*", Elsevier, Amsterdam, 1992.
55. Kaliszan, R., Turowski, M., Buciński, A. and Hartwick, R.A., *Quant. Struct.-Act. Relat.* **14,** 356 (1995).
56. Offord, R.E., *Nature* **211**, 591 (1966).
57. Nyberg, F., Zhu, M.-D., Liao, J.-L. and Hjerten, S., in: "*Electrophoresis '88*", Schafer-Nielsen, C. (Ed.), VCH, Weinheim, 1988.
58. Deyl, Z., Rohlicek, V. and Adam, M., *J. Chromatogr.* **480**, 371 (1989).
59. Rickard, E.C., Strohl, M.M. and Nielsen, R.G., *Anal. Biochem.* **197**, 197 (1991).
60. Survay, M.A., Goodall, D.M., Wren, S.A.C. and Rowe, R.C., *J. Chromatogr.* **636**, 81 (1993).
61. Grossman, P.D., Colburn, J.C. and Lauer, H.H., *Anal. Biochem.* **179**, 28 (1989).
62. Grossman, P.D. and Soane, D.S., *Anal. Chem.* **62**, 1592 (1990).
63. Lukkari, P., Vuorela, H. and Riekkola, M.-L., *J. Chromatogr.* **655**, 317 (1993).
64. Siren, H., Jumppanen, J.H., Manninen, K. and Riekkola, M.-L., *Electrophoresis* **15**, 779 (1994).
65. Jumppanen, J.H. and Riekkola, M.-L., *Anal. Chem.* **67**, 1060 (1995).

6 Chemometric Methodology of QSRR Studies

6.1. Basis, Aims and Strategy

Chemometrics is the chemical discipline which aims to provide maximum information through analysis of chemical data. When computers became more widely available in 1960s, chemometrical approaches to problem solving in chemistry became possible and new research areas were created within the science, one of these being QSRR. Basically, the research methodology now applied in QSRR studies was transferred from studies of quantitative structure-biological activity relationships (QSAR) [1]. It remains similar but, naturally, the emphasis in QSRR is stressed on specifically chromatographic aspects of structure-property relationships, even if one has eventual application in QSAR in mind.

As soon as chromatography came into wider use, smooth changes of retention with carbon number were found for series of homologues analysed at fixed mobile and stationary phase conditions. However, for a more diverse sets of solutes, relationships between retention and empirical or theoretical molecular structural parameters have usually not been unequivocal and required statistical evaluation in order to check the significance of the resulting correlations. At present appropriate statistical calculation computer programs suitable for QSRR are available in every chemistry laboratory. However, these programs are sometimes used in a "black box" manner and the results obtained (and reported occasionally) happen to be insignificant or incorrectly interpreted.

QSRR are derived statistically. There should, however, be an identifiable physical sense behind them. As a matter of fact, one can argue that good retention "prediction" proves the validity of given QSRR and that one should try to discover the physical sense hidden, for example, in an effective structural variable. Certainly, one cannot exclude the possibility that trial and error attempts will result in the identification of some new meaningful descriptors which can easily be calculated, *e.g.*, from structural formulae. The danger is that, in striving for that modern philosopher's stone, one occasionally tends to play a number game. Besides, it should be remembered that if large number of parameters is tested even random numbers may show correlations [2].

It is impossible to discuss here individual calculation methods applied in chemometrics. Professional literature [3–8] and software manuals are recommended.

When starting QSRR study one should at first prepare (select) representative set of reliable chromatographic retention data. Then, a set of well defined structural descriptors should be tested which account for differences in molecular bulkiness, electronic properties and solute shape. Trends should be noted which parameters provide significant input to retention description. The descriptors can next be modified to improve the QSRR. Various *ad hoc* designed structural descriptors can then be tested. If a few evident outliers are identified they have to be analysed both from the view point of reliability of their chromatographic data and inadequacy of the applied descriptors to account for their distinguishable structural features. The resulting QSRR should be interpretable in physical terms. Observing all the statistical rules and recommendations one should select the minimum number of structural descriptors needed to produce QSRR of good predictive ability.

6.2. Multiple Regression Analysis

According to the concept of LFER, a given chromatographic retention parameter may quantitatively (statistically) be described by a set of experimentally or theoretically derived solute structural descriptors

$$\text{Retention parameter} = f(a_1 x_1, \ldots, a_n x_m). \tag{6.1}$$

The coefficients a_1–a_n at the individual m descriptors are calculated by multiple regression. A special case, corresponding to $m = 1$, is the simple regression model $y = a_0 + a_1 x_1$.

There are numerous computer programs available to derive regression coefficients and to evaluate the statistical value of the regression model assumed. There also the computer assisted procedures of identification of the best regression model (at least from the statistical point of view) for given regression values and a set of regressors. A calculation procedure called step-wise regression selects regression model obeying fixed borderline conditions.

For any set of variables which can be assigned to a series of objects, regression equations can be derived. Whether or not is any of the possible equations statistically significant is judged based on several statistical significance parameters. Of these parameters the most often analysed (and reported) ones are denoted by symbols R (r), s, F, t, r_{ij}.

R (or r for simple linear regression) is the correlation coefficient. Its square R^2 (r^2) is the coefficient of determination or the percentage of data variance accounted for by the corresponding regression equation.

The s is the standard error of estimate. It is also called the standard deviation from regression equation. Advanced programs yield additionally standard deviations of individual regression coefficients.

F is a statistical parameter commonly employed for assessing the overall significance of the derived regression equation. It can be calculated by the formula

$$F = (n - k - 1)R^2/[(1 - R^2)k], \tag{6.2}$$

where R is the correlation coefficient, n is the number of data points (regressand or dependent variable values) used to derive the regression equation and k is the number of descriptors (regressors or independent variables) present in the equation. The F values are compared with reference values for given number of degrees of freedom. In effect, significance levels p are determined for the regression equation as well as for individual regressors.

The t-test values for individual regressors quantify their relative statistical significance for the overall correlation provided by the whole regression equation. These t values, along with the standard deviations, serve to calculate the confidence intervals (usually 95%) for individual regression coefficients.

The r_{ij} are the cross-correlation coefficients between the independent variables in the equation. These coefficients must be low to assure the independence of the variables.

To derive good QSRR by multiple regression, the experience and chemical intuition of the researcher are of much help. First of all, one cannot rely solely on statistics. The statistically derived equations should make good chemical sense. A potential for chance correlations when to many variables are surveyed to correlate too few data has already been mentioned [2]. There is also a rule of thumb that 4–6 data points of the regressand should fall per regressor, *i.e.*, if one has retention parameters for 15 solutes the number of regressors (structural descriptors) in the regression equation should not exceed three.

The Committee on Minimal Standards for Reporting Results of Regression Analysis of the International Group for Correlation Analysis in Organic Chemistry published [9] an announcement entitled "*Recommendations for Reporting the Results of Correlation Analysis in Chemistry using Regression Analysis*". The following requirements should be met accordingly.

1. All of the data points (*i.e.*, values of the dependent variable, the quantity to be correlated) in a set should be reported either in the paper or the supplementary material. If values are excluded from the correlation this should be clearly indicated and the reason for exclusion stated.
2. All of the independent variables (*i.e.*, explanatory variables included in the correlation analysis at any stage should be reported even if they are not present in the final correlation.

This requirement is meant to apply to multiple linear regression analyses in which no *a priori* model is available and correlation is carried out with a collection of independent variables some of which are dropped when they fail to contribute significantly to the correlation equation. Values of the independent variables used in the correlation analysis should be given in the paper or in the supplementary material whenever possible. Otherwise their sources should be reported. If values are estimated the method of estimation and its source should be given.

3. Statistics reported should include at least the following:
   a. Standard error of estimate.
   b. Standard errors of the regression coefficients.
   c. At least two of the following:
      (i) R or $100R^2$ (or for simple linear regression r or $100r^2$).
      (ii) F (with confidence levels).
      (iii) f or ψ. The former is intended for use with linear regression through the origin, the latter for linear regression with an intercept.
   d. A correlation matrix or table of correlation coefficients r_{ij} for the correlation of the i-th independent variable with j-th independent variable. These are called zeroth order correlation coefficients.

Optional but recommended by the Committee are the following:

1. Values of R_i for the correlation of the i-th independent variable with all of the other independent variables. In the case of four independent variables, for example, the R_i are $R_{1,234}$, $R_{2,134}$, $R_{3,124}$ and $R_{4,123}$. Values of R for the i-th independent variable on all other possible combinations of the other independent variables are also of interest. The R_i are sufficient, however, for the detection of collinearity among the independent variables.
2. Graphs:
   a. In simple regression a plot of y *versus* x
   b. In multiple regression a plot of y calculated *versus* y observed. Alternatively, a residual plot may be used. A plot of the first two principal components of the x matrix and information about the variance matrix and information about the variance explained would also be of use. The number of degrees of freedom should be given on the graph or in its caption.

6.3. Multivariate Methods of Data Analysis

The multivariate methods of data analysis, like discriminant analysis, factor analysis and principal component analysis, are often employed in chemometrics. In chemical terms the objective may be interpretation of physical meaning of the data analysed, prediction of data for new chemical objects and objects classification. Widest popularity in QSRR studies got the technique of principal component analysis (PCA). By PCA one reduces the number of variables in a data set by finding linear combinations of those variables that explain most of the variability. Factor analysis (FA) is similar to PCA except that factor weights are scaled so that their sum of squares is equal to the associated eigenvalue of the correlation or covariance matrix. Thus, the sum of square of the scaled factor weights is related to the total variance explained by a given factor. Having selected the number of factors to be retained in final solution one can rotate the factor matrix. The methods of rotation usually reported are called: varimax, equimax and quadrimax. Rotated factors are often easier to interprete.

The PCA allows one to find the principal components given either the original variables or a correlation or covariance matrix. Coefficients of each principal component are determined by first computing the eigenvalues of the covariance matrix or the correlation matrix. Standard commercially available softwares tabulate the eigenvalues, the proportion of the total variance accounted for by each component, the component weights and the values of individual principal components. Especially convenient in QSRR are plots of component weights for the first two principal components for each variable. Analogously, the scatterplots for the first two principal components illustrate distribution of objects according to their principal components. Biplots account simultaneously for distribution of both variables and principal components [10,11].

Factorial methods of data analysis are normally used in chemistry if multiple regression methods fail. For example, reliability of multiple regression may be questioned if there are significant intercorrelations among the regressors. The cross correlation between the independent variables can be assessed from the correlation matrix of the variables. If the dependent variable is included, the variance-covariance matrix is constructed. Such matrices may be transformed by the methods of linear algebra into new matrices containing non-zero elements only on the diagonal. These are the so-called eigenvalues of the matrix. With each eigenvalue of the matrix thus obtained, an eigenvector is associated which is a linear combination of the original set of variables. The characteristic feature of the eigenvectors is that the correlation coefficient between them is exactly zero. This means that eigenvectors are completely orthogonal, as opposed to the original set of variables. If there is significant covariance in the set of original variables, most of the total variance will be accounted for by a number of eigenvectors equal to a fraction of the original number of variables. A reduced set containing only the major eigenvectors (principal components) may then be examined or used in various ways [12,13].

The determination of how many eigenvectors may be reasonably ignored in PCA is a subject of discussion [8,14,15]. In any case, if only the principal components are considered, new orthogonal variables can be constructed from the eigenvectors, and hence the dimensionality of the parameter space can be reduced while most of the information in the original variable set is retained.

In QSRR studies one usually attempts to assign a physicochemical meaning to the individual abstract factors derived mathematically. Sometimes, the multiple regression analysis is applied to this end with various structural descriptors as independent variables.

Factor analysis can also be employed to detect multicolinearity among independent variables to be applied in multiple regression analysis [16]. Multicolinearity is the correlation which involves more than two regressors. Whereas simple correlations between two regressors (structural descriptors) can easily be found in correlation matrix, the multicolinearity cannot be readily seen if large number of regressors are considered. In such cases a mathematical statistics rule [17] is applied that if any individual eigenvalues of the covariance matrix are small or the sum of their reciprocals is greater than five times the number of regressors, it can be said that the variables are intercorrelated.

There are modifications of standard factorial methods of data analysis reported in QSRR studies. The correspondence factor analysis (CFA) has been popularised by Chrétien and co-workers [18,19]. There is also an approach in which principal components are regressed against retention data in a multiple linear regression, *i.e.*, principal component regression (PCR). The partial least square (PLS) approach with cross-validation was developed by Wold and co-workers [20,21].

The multiple linear regression (MLR), principal component regression (PCR) and partial least squares (PLS) methods have been compared [22–24]. No method can be regarded as the best choice: all have their advantages and disadvantages. The choice of a given method depends on prefered criteria of optimization. If the optimization criteria are focused only on maximizing the correlation of x variables with y responses usually MLR is chosen. If one is interested in maximizing the variance explained by the x variables with the factors then PCR is applied. If one wishes to do both to the same degree simultaneously then PLS should be chosen [25].

The Committee on Minimal Standards for Reporting the Results of Principal Component Analysis of the International Group for Correlation Analysis in Organic Chemistry [26] published in 1990 "*Recommendations for Reporting the Results of Principal Component Analysis in the Field of Quantitative Structure Activity Relationships*" [27].

Regarding reporting the results of PCA it is suggested that the covariance or the correlation matrices may be included if they are "of moderate size". Otherwise a comprehensive description of the most interesting correlations should be given.

The criteria for the choice of a certain number of components should be clearly indicated. The percent of the explained sum of squares should be given successively for one, two, *etc.*, components. If the data are not scaled, the standard deviation (in the original units) is one of the most telling characteristics. The degrees of freedom should be stated.

The Committee observes that the explanation of a significant amount of variation in poorly structured (correlated) data necessitates a large number of factors. In this case, the analysis is probably not effective.

It is recommended that the correlation coefficients of the variables with each of the principal factors be given. The factors represent the weight of the variable in the definition of the factorial space.

Any type of rotation or target transformation should be clearly indicated, as well as the criteria for using such a transformation.

Plots of the data points (scores) in two- (or three-) dimensional space defined by 2 (or 3) principal factors are recommended as useful visual aids. Projections in the plane defined by the first two factors are usually used, even when the dimensionality is larger than 2. Projections in planes defined by other pairs of factors may also be pertinent.

The Committee also recommends basic textbooks and reviews [8,12,14,15,28,29].

Regarding the sources of data for PCA the Committee states that if only a few literature source (easily accessible) are used, the bibliographic references suffice. Reproducing the data is recommended if numerous sources are used. Large data tables may be supplied as supplementary material if this is the policy of the journal concerned. Whenever possible, it is advisable to report (or give an estimate of) uncertainties in the data.

There are the following remarks of the Committee as regards to preprocessing. They state that PCA can use either a full data matrix or an incomplete one [8]. If the matrix used is a submatrix of a large one, the criteria for the choice of a particular subset should be given.

Estimated data should be clearly indicated. If missing data are calculated by an algorithm, or any other means, the method used should be stated. If an incomplete matrix is used, the number of missing items or the filling rate of the matrix in percent should be given. If some data are deleted for *a posteriori* reasons as outliers after a preliminary treatment, this should always be stated and the criteria for deletion indicated.

As far as the statistical treatment of data is concerned the Committee suggests what follows. First, the fundamental technique should be stated clearly (*e.g.*, covariance matrix technique). A brief statement about the type of software may be useful. Second, the treatment of raw or transformed data may be prefered according to the main purpose and with respect to the character of the data (physical quantity, unit, magnitude, variance, uncertainties). Data are usually standardized within columns, by subtracting the mean from each term (centering) and by dividing by the standard deviation (scaling). Other standardization methods [12,25] may also be used. It is essential to indicate clearly any method of standardization of data.

6.4. Neural Network Analysis

Neural networks (NN) is a method of analysis which is to emulate the brain's way of working. The method has recently got into a wider use. However, already in November 1992 Chemical Abstract Services listed nearly 500 articles published under the code word "neural networks" which were related to chemistry [30].

In general the NN are considered modern powerful tools and techniques for doing signal processing, modeling, forecasting and pattern recognition. According to some users, the NN may replace conventional statistical techniques, pattern recognition and classification methods, and all the methods using linear regression, polynomial regression and curve fitting techniques.

Neural network has its input neurons which load the system with descriptor values. Next, there are the hidden layers which weight and mix the incoming signals, and an output layer with neurons predicting the calibrated response values [30–32]. The system is taught by an interactive algorithm which adjusts the weights, w_{ij}, of the incoming signals, Inp_{ij}, at each neuron j. All the input signals i to a neuron j are described by the formula

$$\mathrm{Net}_j = \sum_{i=1}^{j} (\mathrm{In}\, p_{ij} \cdot w_{ij}). \tag{6.3}$$

The nonlinear transformation occurs at every neuron. The output, Out_j, from every neuron is some nonlinear transformation such as a sigmoid function (Eq. (6.4)), a threshold logic function (Eq. (6.5)), a quadratic function (Eq. (6.6)) and a hyperbolic tangent function (Eq. (6.7)):

$$\mathrm{Out}_j = \{1 + \exp[-(b_{0,j} + b_{1,j}\ \mathrm{Net}_j)]\}^{-1}, \tag{6.4}$$

$$\mathrm{Out}_j = \max[(0, \min(1, b_{0,j} + b_{1,j}\ \mathrm{Net}_j)], \tag{6.5}$$

$$\mathrm{Out}_j = b_{0,j} + b_{1,j}\ \mathrm{Net}_j + b_{2,j}\ \mathrm{Net}_j^2, \tag{6.6}$$

$$\mathrm{Out}_j = \tanh(b_{0,j} + b_{1,j}\ \mathrm{Net}_j). \tag{6.7}$$

Neural networks predictive power has been compared with other chemometric methods of data analysis, namely the multiple linear regression [33–35] as well as with principal component regression and partial least squares [36,37]. Some authors report better performance of NN. For example in case of chromatography Peterson [34] found NN significantly better than MLR at modeling gas chromatographic retention data. However, Cupid *et al.* [35] found the NN approach "as successful as the regression analysis" in predicting HPLC retention of substituted aromatic solutes.

References

1. Hansch, C. and Fujita, T., *J. Am. Chem. Soc.* **86**, 1616 (1964).
2. Topliss, J.G. and Edwards, R.P., *J. Med. Chem.* **22**, 1238 (1979).
3. Shorter, J., "*Correlation Analysis in Organic Chemistry: An Introduction to Linear Free-Energy Relationships*", Clarendon, Oxford, 1973.
4. Massart, D.L., Vandeginste, B.G.M., Deming, S.N., Michotte, Y. and Kaufman, L., "*Chemometrics: A Textbook*", Elsevier, Amsterdam, 1988.
5. Brereton, R.G., "*Chemometrics. Application of Mathematics and Statistics to Laboratory Systems*", Ellis Harwood, New York, 1990.
6. Brereton R.G. (Ed.), "*Multivariate Pattern Recognition in Chemometrics*", Elsevier, Amsterdam, 1992.
7. Deming, S.N. and Morgan, S.L., "*Experimental Design: A Chemometric Approach*", Elsevier, Amsterdam, 1993.
8. Wold, S., Esbensen, K. and Geladi, P., *Chemom. Intell. Lab. Syst.* **2**, 37 (1987).
9. Charton, M., Clementi, S., Ehrenson, S., Exner, O., Shorter, J. and Wold, S., *Quant. Struct.-Act. Relat.* **4**, 29 (1985).
10. *Statistical Procedures Reference Manual*, Statgraphics® Plus, Manugistics, Rockville, Maryland, 1992.
11. Johnson, R.A. and Wichern, D., "*Applied Multivariate Statistical Analysis*", Prentice-Hall, New Jersey, 1982.
12. Malinowski, E.R. and Howery, G.D., "*Factor Analysis in Chemistry*", Wiley, New York, 1980.
13. Cramer, R.D.III, *J. Am. Chem. Soc.* **102**, 1837 (1980).
14. Mardia, K.V., Kent, J.T. and Bibby, J.M., "*Multivariate Analysis*", Academic Press, London, 1979.
15. Jolliffe, I.T., "*Principal Component Analysis*", Springer-Verlag, New York, 1986.
16. Hasan, M.N. and Jurs, P.C., *Anal. Chem.* **55**, 263 (1983).
17. Chatterjee, S. and Price, B., "*Regression Analysis by Example*", Wiley, New York, 1977.
18. Hirsch, R.F., Gaydosh, R.J. and Chrétien, J.R., *Anal. Chem.* **52**, 723 (1980).
19. Walczak, B., Dreux, M. and Chrétien, J.R., *Chromatographia* **31**, 575 (1991).
20. Wold, S., Eriksson, L., Hellberg, S., Jonsson, J., Sjöström, B., Skageberg, B. and Wikstrom, C., *Can. J. Chem.* **66**, 1814 (1987).
21. Eriksson, L., Jonsson, J., Sjöström, M. and Wold, S., *Quant. Struct-Act. Relat.* **7**, 144 (1988).
22. Kowalski, B.G., *Chemom. Intell. Lab. Syst.* **9**, 177 (1990).
23. Thomas, E.V. and Haaland, D.M., *Anal. Chem.* **62**, 1091 (1990).
24. Martens, H. and Naes, T., "*Multivariate Calibration*", Wiley, Chichester, 1991.

25. Frank, I.E., *Chemom. Intell Lab. Syst.* **8**, 109 (1990).
26. An associated organization of the International Union of Pure and Applied Chemistry.
27. Gal, J.-F., Maria, P.-C., Chastrette, M., Zakarya, D., Exner, O., Haldna, Ü., Sjöström, M., Wold, S. and Zalewski, R.I., *Quant. Struct.-Act. Relat.* **9**, 132 (1990).
28. Dagnelie, P., "*Analyse Statistique à Plusieurs Variables*", Les Presses Agronomiques de Gembloux, Gembloux, 1982.
29. Massart, D. C. and Kaufman, L., "*The Interpretation of Analytical Chemical Data by the Use of Cluster Analysis*", Wiley, New York, 1983.
30. Zupan, J. and Gasteiger, J., "*Neural Networks for Chemists. An Introduction*", VCH, Weinheim 1993.
31. Janssen, P.A., *Anal. Chem.* **63**, 357A (1991).
32. Gemperline, P.J., *Chemom. Intell. Lab. Syst.* **15**, 115 (1992).
33. Aoyama, T., Suzuki, Y. and Ichikawa, H., *J. Med. Chem.* **33**, 2583 (1990).
34. Peterson, K.L., *Anal. Chem.* **64**, 379 (1992).
35. Cupid, B.C., Nicholson, J.K., Davis, P., Ruane, R.J., Wilson, I.D., Glen, R.C., Rose, V.S., Beddell, C.R. and Lindon, J.C., *Chromatographia* **37**, 241 (1993).
36. Gemperline, P.J., Long, J.R. and Gregoriou, V.G., *Anal. Chem.* **63**, 2313 (1991).
37. Borggaard, C. and Thodberg, H.H., *Anal. Chem.* **64**, 545 (1992).

7 Simple Linear Structure-Retention Relationships

7.1. General Remarks

The scope of this book is limited to the effects of structure changing within a series of solutes on retention of individual members of the series, provided that all the solutes tested are chromatographed at the same conditions or the retention data concerned have been normalized. This means that author's interest is not extended to the effects of changes in nature and composition of mobile phase and/or stationary phase on separation. In other words the QSRR are not derived and discussed here in order to optimize chromatographic separations by proper choice (design) of chromatographic systems. Of course, indications regarding optimization may result from QSRR analysis. However, readers aiming at determining the optimum chromatographic conditions for solving a defined separation task should at first study literature regarding the chromatographic method development (at present also relying on chemometrics) (*e.g.*, refs. [1–5]).

7.2. Retention of Homologues

Linear relationships between retention parameters and carbon number for homologous series of solutes were observed since long. The slopes of respective lines for various homologous series chromatographed at the same conditions were reported to be very similar [6,7]. This observation is known under the name of Martin equation and is considered valid for many forms of chromatography:

$$\log K\,(\text{or}\ \log k') = A + Bn, \tag{7.1}$$

where K is partition coefficient, k' is capacity factor, n denotes the homologue number and A and B are constants. On this relationship is based the Kovâts retention index in gas chromatography as well as the HPLC retention prediction schemes of Smith [8], Jandera [9], Gilpin [10] and Carr *et al.* [11]. Also the UNIFAC approach [12] to predicting transfer of various homologues from aqueous to organic systems is based on Martin equation [13]. There are recent reports, however, that in reversed-phase HPLC on medium and long-chain monofunctional bonded silicas having a high bonded density the $\log k'$ *versus* carbon number plots are often not simple straight lines, particularly when organic-rich mobile phases are used [14,15]. In such cases the relationships are curvilinear or may be considered to consist of (at least) two segments of straight lines.

The analysis of dependence of retention on carbon number of solutes in terms of what is now called QSRR has long history. Already in 1973 Ladon and Sandler [16] derived regression equation relating the solute carbon number to its gas chromatographic retention time on three stationary phases. These equations hold true for classes of solutes containing the same kind and number of functional groups, such as alkanes, cycloalkanes, alkenes, cycloalkenes and alkadienes. However, the position of the hydroxyl group in primary and secondary straight-chain alcohols was found to affect the retention of homologous series of that class of solutes [17]. If the solutes were not homologous but otherwise closely related the plots of Kováts indices *versus* carbon number remained linear [18]. The relationships between carbon number and GC retention of homologues have extensively been studied from the point of view of the reliability of the retention index system. Interested readers are referred to reports by Smith *et al.* [19] and Budahegyi *et al.* [20].

The plots of GC retention index against carbon number of homologues given by Haken and Karhonen [21] illustrate that starting from chain length of four to five carbon atoms the retention index increments for methylene group on a given stationary phase determined at a given temperature are constant. This is not the case, however, for the first three members of the series.

Various attempts were undertaken to account for non-linearities often observed in retention-carbon number of homologues plots. Some appear formally interesting but questionable after deeper inspection. For example, for estimation of Kováts indices of homologues Golovnya and Grigoreva [22] proposed the following equation:

$$I^{\mathrm{St.phase}(T)}{}_{R_nX} = \alpha + \beta_n + \gamma \frac{\log n}{n} + \frac{\delta}{(n-2)^2 + 0.1}, \tag{7.2}$$

where $I^{\mathrm{St.phase}(T)}{}_{R_nX}$ is the Kováts index of a solute with n carbons in the aliphatic chain belonging to a homologous group with a functionality X chromatographed at temperature T on a given stationary phase; α, β, γ and δ are experimentally determined coefficients. For instance, for amines of the series R_nNH_2 chromatographed at 100°C on *Apiezon L*, the coefficients α, β, γ and δ were 281.09, 97.97, −213.66 and 1.31, respectively. For the same solutes chromatographed on *Carbowax* 1000 the corresponding values were 623.18, 98.25, −409.03 and 1.02. Unfortunately, the statistical significance of the last two terms on right side of Eq. (7.2) has not been proved. In any case, the last term is evidently meaningless, except perhaps for carbon number 2.

Linear relationships of Kováts index on carbon number of homologues may be found in literature for various organic chemical families. They were reported also for organosilicon derivatives [23,24]. It is practically impossible to list all the publications reporting regular changes of GC retention with carbon number in homologous series. In general, the linearity starts from carbon number 4. All these publications confirm the rule established in the fundamental work of James and Martin [25]. As one would expect, similarly to carbon number, linear correlations between GC retention parameters and number of other atoms have been found, *e.g.*, chlorine, silicon, sulfur, nitrogen and oxygen atoms as well as the number of phenyl groups and double bonds [26,27].

Recently, Wang *et al.* [28] reported a linear relationship between carbon number and retention parameter of homologous hydrocarbons in supercritical fluid chromatography. They found a common intersection point among the $\ln k'$ *versus* n straight lines of homologues for different temperatures.

It should be noted that excellent linearity of retention-carbon number relationships in GC is generally reported for nonpolar solutes chromatographed on nonpolar stationary phases. Situation is not that obvious when retention data are compared for a series of solutes of varying polarity chromatographed on polar stationary phases. A good recent example of deviations from linearity observed for such systems is by Wasiak [29] who reported GC retention of nitroalkanes on Cu(II) and Ni(II) acetyloacetonates bonded to silica surfaces to solute carbon number.

Linear relationships between retention data and carbon number have also been reported for various forms of liquid chromatography. Most extensively these relationships have been and still are studied in case of reversed-phase liquid chromatography (RP LC). Examples of early systematic studies of this kind regarding RP paper chromatography and RP TLC may be the works by Green *et al.* [30] and Clifford *et al.* [31]. Deviations from linearity of R_M from RP TLC against n plots was reported by Boyce and Milborrow for aliphatic compounds [32]. In cation exchange chromatography curvilinear plots were reported by Murakami [33].

Relationships between $\log k'$ from reversed-phase HPLC and carbon number of closely related structurally solutes are usually reported to be linear at some limited range of the aliphatic chain length [34,35]. The range of linearity does normally not exceed six to eight carbon atoms, depending on the functional groups present in the test series. Apparent non-linearity of $\log k'$ *versus* carbon number plot for three groups of substituted alkylbenzoates is illustrated in Fig. 7.1 after Tomlinson *et al.* [36].

Character of the relationship between $\log k'$ and carbon number of homologous solutes is analysed in studies on the properties of HPLC stationary phases. It has been demonstrated that with monofunctional bonded silica stationary phases having a high bonding density the $\log k'$ *versus* n plots are curvilinear, particularly if organic-rich mobile phases are used [14]. In case of polyfunctional bonded silicas and polymeric materials such as polyvinylbenzenepolystyrene (either non-bonded or bonded by *n*-octadecyl moieties), the plots of $\log k'$ *versus* n of homologous solutes are sigmoidal (Fig. 7.2).

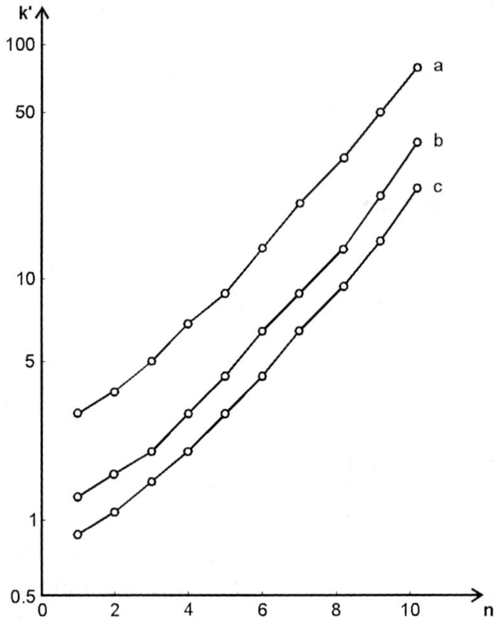

Figure 7.1. Relationship between logarithm of capacity factor, *k'* and carbon number of alkyl chain, *n*, for homologous series of (a) 3-Cl–; (b) 2-OCH_3– and (c) 2-NO_2– substituted alkylbenzoates. (Reprinted from Tomlinson, E., Poppe, H. and Kraaak, J.C., *Int. J. Pharm.* 7, 225, 1981 with kind permission of Elsevier Science-NL, Sara Burgerhartstraat 25, 1055 KV Amsterdam, The Netherlands.)

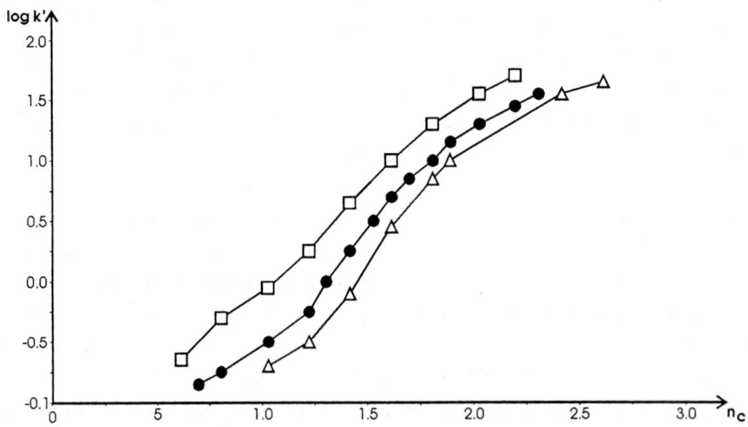

Figure 7.2. Plots of log *k' versus* carbon number *n* of solute for HPLC on a polyfunctional C_{18}-bonded silica (Supelcosil LC PAH) with pure methanol eluent at 25°C. (□) *n*-alkanes, (•) methyl-*n*-alkanoates, (Δ) *n*-alkanols. (Reprinted from Tchapla, A., Heron, S., Lesellier, E. and Colin, H., *J. Chromatogr. A* 656, 81, 1993 with kind permission of Elsevier Science-NL, Sara Burgerhartstraat 25, 1055 KV Amsterdam, The Netherlands.)

7.3. Simple Correlations of Retention with Molecular Bulkiness Descriptors

There are a few publications on linear correlation between retention parameters and molecular mass observed for solutes that are not homologues [37]. Interesting appears the proportionality between GC retention volume and a quantity $(M_w - 170)^{1/3}$, where M_w is molecular mass [38]. This relationship was found for organoarsenic compounds chromatographed on a nonpolar silicone phase. In the same paper [38] retention was correlated to parachor. Parachor was also reported to correlate to GC retention data of related solutes by Wurst [39].

Molecular refractivity and polarizability appeared the more effective than molecular mass descriptors of retention of nonhomologous (but closely related) solutes. With molecular refractivity as independent variable correlations of GC retention indices were reported for organometallic solutes [40,41]. There also are other early examples of such correlations for several classes of organic substances chromatographed on nonpolar stationary phases [42–44].

Good correlations between GC retention indices determined on nonpolar stationary phases and molar refractivity for various alkyl derivatives of both phenylpropionic and phenylacetic acid were found by Macek and Smolkova-Kuelemansová [45]. Other group of Czech workers [46] analysed GC data of variously substituted dibenzo[*b*, *f*]-thiepins. Kováts index determined on a nonpolar phase correlated significantly to molar refractivity of solutes ($r = 0.845$). In case of a more congeneric set of arylkoxy and alkoxyarylaliphatic acids the correlation was obviously much higher ($r = 0.941$) [47].

Significant correlations between molecular polarizabilities and GC retention data of closely related but not homologous compounds have also been reported [48–51].

Van der Waals volume has been another bulkiness descriptor often correlated to retention. Wurst and Churacek [52] demonstrated linear relationship between GC retention data and molar volume of a series of organosilicon compounds. Such relationships were also reported by Saura Calixto and Garcia Raso [53]. For 61 alkylbenzenes correlations with GC retention indices obtained on *Squalane* were higher in case of Van der Waals volume ($r = 0.94$) than in case of molar refractivity ($r = 0.92$) [54]. Respective relationship derived for Kováts indices determined on a polar phase (*Carbowax 20M*) were lower ($r = 0.88$).

An application of Van der Waals volumes in QSRR regarding HPLC capacity factors was described by Jinno and Kavasaki [55]. These authors used three stationary reversed-phase materials of increasing alkyl chain length to generate retention data of 19 isomeric alkylbenzenes. Significant correlation between $\log k'$ and Van der Waals volume were found, especially in the case of C_8 and C_{18} phases. In another paper the same authors correlated $\log k'$ determined on various reversed-phase columns for a set of polycyclic aromatic hydrocarbons (PAH) with several structural descriptors, including Van der Waals volume [56]. As expected for such closely related solutes like PAH, high correlation was found between $\log k'$ and Van der Waals volume as well as when any other molecular size related quantity was used as independent variable.

Studies of Jinno and Kawasaki [55–57] demonstrated that Van der Waals volume as well as other solute bulkiness descriptors correlate well with $\log k'$ from RP HPLC of closely related nonpolar solutes, like PAH or alkylbenzenes. However, if more diverse sets of solutes are considered, correlations become insignificant. Thus, for PAH chromatographed on a C_{18} stationary phase with methanol-water (75:25 v/v) eluent, the $\log k'$ *versus* Van der Waals volume correlation is high ($r = 0.989$). If Van der Waals surface area is the independent variable, an analogous correlation is $r = 0.982$. If alkylbenzenes are chromatographed at identical conditions, then for $\log k'$ *versus* Van der Waals volume the correlation is $r = 0.992$, and similarly for $\log k'$ *versus* Van der Waals surface $r = 0.991$. However, if the solutes chromatographed at identical conditions are benzene derivatives possessing such functional groups like NH_2, OH, CN, NO_2, $COOCH_3$, *etc.*, the respective correlations drop to $r = 0.35$ for $\log k'$ *versus* Van der Waals volume and to $r = 0.34$ for $\log k'$ *versus* Van der Waals surface [56]. For a larger set of substituted benzene derivatives chromatographed in a C_{18} RP HPLC system with a water-acetonitrile mobile phase, the variables $\log k'$ and Van der Waals volume are completely independent ($r = 0.08$). Similarly, correlations with other bulkiness descriptors are low [57].

Hanai *et al.* [58] determined $\log k'$ in RP HPLC for several groups of solutes at six temperatures. Next, enthalpy changes ΔH for the solutes were calculated. Smooth curvilinear relationships were found between ΔH and Van der Waals volume for individual families of structurally related solutes.

Literature retention data of barbituric acid derivatives determined by various chromatographic methods were studied by Bojarski and Ekiert [59]. The correlations of Van der Waals volumes with TLC, GC and HPLC retention parameters have been significant, especially in the case of GC and TLC data considered. As expected, the significance level of the relationships decreased with increasing diversity of structure of the solutes.

GC retention data determined on nonpolar stationary phases and RP HPLC capacity factors for a group of phenylacetic and phenylpropionic acids were related to Van der Waals volumes of the solutes [45]. It has been found that in RP HPLC Van der Waals volumes gave lower correlation with retention parameters than in GC. The correlation for GC data determined on less polar stationary phase was better than for the data determined on a more polar column.

Reported simple correlations of retention parameters and solute bulkiness descriptors appear significant for closely related series of solutes. The lower is the polarity of a chromatographic system (stationary phase), the better are the correlations. Regarding the structure of solutes one can observe that simple description of retention in terms of a bulkiness parameter is obtained if either the solutes analysed are nonpolar (*e.g.*, aliphatic hydrocarbons) or are of closely similar polarity (*e.g.*, congeners possessing the same functional group). In such a situation, differences in retention among solutes can be assigned to the differences in their nonspecific (dispersive) interactions with the components of a chromatographic system. Thus, bulkiness descriptors can account for separation of solutes when dispersive interactions are the only interactions effective in a given chromatographic system or when differences in polar interactions among solutes are not significant.

Of a number of solute bulkiness descriptors studied in simple QSRR, Van der Waals volume and molar refractivity appeared the most appropriate. As it will be demonstrated latter on basis of multiparameter QSRR the total energy, Van der

Waals surface area and water accessible surface area have proved to be even more effective discriminators of dispersive properties of solutes. To that category of descriptors seem to belong some topological indices discussed separately in Section 7.5.

7.4. Simple Correlations of Retention with Molecular Polarity Descriptors

The ability of solute molecule to participate in dispersive intermolecular interactions may somehow be approximated by bulkiness descriptors. However, the ability of solutes to take part in what is more or less intuitively called polar interactions, is difficult to be characterized by means of a one-dimensional scale. Thus, simple QSRR involving a single solute polarity descriptor are rare.

Naturally, the first polarity parameter tested from the point of view of retention prediction has been the dipole moment. Already in 1956 James [60] related the chromatographic behaviour of a series of isomeric xilidines to their dipole moment. Correlations of GC retention indices with dipole moments for some organosilicon compounds were also reported [61,62]. Square of dipole moment was found to correlate to GC retention of aromatic mononitro derivatives but not to retention of dinitro derivatives [63]. Also, the dielectric constant of solutes was correlated to GC retention assuming that it can reflect the ability of solute to take part in dipole orientation interactions [64].

To prove unequivocally a quantitative relationship of retention on dipole moment it would be best to have a series of solutes of the same size, shape, hydrogen bonding capacity, *etc*., but differing in their dipole moments. There is a report on GC behaviour of such a series of 8 solutes, namely methyl-X-ketononanoates, where X denotes varying position of keto group [65]. Linear plot of logarithm of retention time against position of keto group can be interpreted as a retention-dipole moment relationship.

Normally, changes in dipole moment within a series of solutes are accompanied by changes in other polar and nonpolar properties. Although total dipole moment is still occasionally reported as a significant solute structure descriptor in QSRR [66] it cannot account for all the specific polar features of the compounds. Uncertainty regarding actual value of dipole moment of solute in specific chromatographic environment makes the problem even more complex.

Limited value of dipole moment for polarity description has already been indicated in the pre-QSRR time by Scott [67] and by Karger *et al.* [68]. These authors demonstrated that local dipoles, rather than the overall molecular dipole moment, contribute to retention. Multiparameter QSRR discussed in Section 7.7 confirm this observation.

7.5. Simple Correlations of Retention with Molecular Shape Descriptors

To describe molecular shape in quantitative terms is even more difficult than polarity. A special difficulty in QSRR studies is that the shape increment to retention is generally of minor importance in comparison to differences resulting from both polar and dispersive interactions. In the case of polar solutes, the molecular shape differences are, at least in part, incorporated in parameters designed to describe the changes in solute polarity.

Importance of solute shape for some chromatographic separations was observed first in GC [69,70]. It appeared that retention of polycyclic aromatic hydrocarbons (PAH) on liquid crystal stationary phases depended on degree of elongation of the solute molecule. This observation led Kaliszan *et al.* [71] to propose a shape descriptor η (see Section 4.4.8.1). A linear relationship was found [72,73] between RP HPLC retention parameter of PAH isomers of molecular weight 278 and the shape parameter. The parameter found application in several multiparameter QSRR as will be discussed in Section 7.8.

7.6. Simple Correlations of Retention with Molecular Graph Derived Descriptors

Introducing his connectivity index $^1\chi$ Randić [74] provided linear plots relating this structural descriptor to Kováts indices of several C_5–C_8 branched alkane isomers. Soon after that Karger *et al.* [75] found the molecular connectivity index useful for prediction of RP TLC and RP HPLC retention data of phenols and alcohols containing various normal, branched and cyclic hydrocarbon moieties.

The first regular QSRR study based on regression analysis also employed the first-order connectivity index [76]. In this study, published in 1977, a set of 20 RP TLC R_M data extrapolated to a primary aqueous system was obtained for heterocyclic solutes containing various alkoxy, aryloxy, aminoalkoxy and alkylamine substituents. The correlation between R_M and $^1\chi$ was $r = 0.95$. In the same year 1977 another QSRR study was published [77] concerning GC retention data taken from literature [78] as determined for a series of 36 complex cyclic alcohols and esters on a nonpolar phase (SE-30) and two polar phases (Carbowax 20 M and EGSSX), statistical parameters were determined (Table 7.1) of the regression equation of the form:

$$I = a\,^1\chi + b, \tag{7.3}$$

where I is Kováts index, $^1\chi$ is first-order connectivity index; a and b are regression coefficients.

Table 7.1 Regression coefficients, standard errors of estimate and correlation coefficients of Eq. (7.3) for a group of cyclic alcohols and esters chromatographed on three GC columns [77].

Statistical Parameters	*Stationary phases*		
	SE-30	*Carbowax 20 M*	*EGSSX*
Shape a	148.5	73.5	77.7
Intercept b	397.1	1208	1360
Standard error s	32	74	73
Correlation coefficient r	0.952	0.587	0.623

Statistical parameters given in Table 7.1 demonstrate a good correlation for retention data determined on the nonpolar phase SE-30. The correlation obtained in case of polar stationary phases are of little statistical value, however. Whereas on the nonpolar phase the nonspecific dispersive interactions determine differences in retention among the solutes, the more specific polar interactions become discriminative in the case of polar phases and the polar solutes considered. Thus, the connectivity index should be viewed as a valuable dispersive intermolecular interactions descriptor.

Latter on many QSRR studies have been published with connectivity indices as the GC retention predictors. These studies concern various isomeric, aliphatic, aromatic and polycyclic hydrocarbons [79–86]. Hydrocarbons, as nonpolar compounds, form an especially suitable group of solutes for QSRR analysis employing molecular connectivity indices. Correlations obtained are good at least as long as the solutes of the same degree and position of unsaturation are considered.

Kier and Hall [87] related Kováts indices of separate groups of aliphatic alcohols, ketones, ethers and esters to various connectivity indices. For each group considered separately, good correlations were obtained for retention data determined on a nonpolar phase. However, when the data determined on a polar stationary phase were considered there was no satisfactory description of retention. From studies on chlorinated benzenes Sabljić [88] concluded that for the description of GC retention on a nonpolar stationary phase the first-order molecular connectivity index was a better structural parameter than the Wiener number or the Balaban index.

A high discriminative power of connectivity index towards isomers of methyl esters of higher fatty acids chromatographed on nonpolar GC phases was demonstrated by Kaliszan [89]. The index appeared unable to account for differences in unsaturation of individual subgroups of solutes.

QSRR analysis of RP HPLC retention data in terms of connectivity indices was undertaken by Jinno and Kawasaki [55]. A rather moderate correlation was found for a series of alkylbenzenes chromatographed on C_2, C_8 and C_{18} columns. The octadecylsilica-derived data appeared the least colinear with ${}^1\chi$ ($r = 0.84$). For the two remaining columns r was about 0.94. The same authors [57] reported the lack of correlation ($r = 0.18$) between $\log k'$ from RP HPLC and molecular connectivity for a set of benzene compounds containing common functional groups. Rather moderate, at best, have been the correlations between $\log k'$ from RP HPLC and connectivity indices for a set of 12 *p*-benzoquinones and *p*-hydroquinones [90]. Also, the correlation coefficient of GC retention indices with molecular connectivity for a mixture of alkanes, alkenes, alcohols, ethers, esters, aldehydes and ketones chromatographed on a polar phase (DEGS) have been found [91] to be low ($r = 0.17$).

Of various modifications of connectivity indices the best performing in QSRR seems to be the first-order valence connectivity index ${}^1\chi$ (see Section 4.4.8.2). In support of such an opinion the literature reports regarding nonhomologous series of several solutes may be quoted [92–96]. Specifically, the correlation of GC retention index determined on nonpolar phases for nitrated PAH *versus* ${}^1\chi^v$ was high ($r = 0.970$) [95]. Of a variety of connectivity indices tested, the same descriptor best correlated to retention indices of a series of ketones determined using linear temperature-programmed capillary gas chromatography on a nonpolar stationary phase [96]. From studies of barbiturates Stead *et al.* [93] and Bojarski and Ekiert [94] concluded that different modifications of connectivity indices have no unambiguous advantage over classical $^1\chi^v$. Correlations of various chromatographic data of barbiturates with connectivity indices are good for a closely related subseries of solutes (*e.g.*, when simple saturated aliphatic substituents are present). The more diverse the structure of the solutes considered, the lower the correlation coefficients independent of the chromatographic technique used for generating the retention data. If, for example, QSRR involving $\log k'$ and $^1\chi^v$ are concerned, when a subgroup of aliphatic substituted derivatives have a correlation coefficient that is $r = 0.992$, r drops to 0.966 after inclusion of alkenyl-substituted barbiturates and to 0.914 when *N*-methyl-C-5-disubstituted derivatives are additionally considered ($r = 0.743$) [97].

There are the single-parameter QSRR in which Wiener number W is used as structural descriptor. For example, a relationship has been proposed [83] for description of GC retention index I of a series of monoalkylbenzenes and *ortho*-dialkylbenzenes of the following form:

$$I = (244 \pm 4)W^{(0.297 \pm 0.003)}. \tag{7.4}$$

The authors [83] advocate the superiority of the Wiener number W over the Randić connectivity index χ.

In studies of both standard slow-equilibrium and chromatographic partition coefficients of polycyclic aromatic hydrocarbons and their nitrogen derivatives the index $^1\chi^v$ yielded good QSRR but only if the two groups of solutes were considered separately. However, upon combining both groups, a correlation coefficient of only 0.858 was obtained [98].

For a series of 114 chlorinated dibenzofurans GC retention indices were determined on a nonpolar stationary phase by Hale *et al.* [99]. Original authors derived QSRR describing retention in terms of 9 "dummy" variables representing possible substitution patterns. These data were also used by Dunn *et al.* [100] who derived QSRR in terms of calculated molecular surface area. Later on it was demonstrated [101] that the first-order valence connectivity index $^1\chi^v$ provided better correlation.

To describe the GC retention times for 26 tetra-*n*-alkyl-silanes determined on a nonpolar column Kupchik [102,103] found the index $^1\chi^v$ to be the best regressor of several topological indices tested in a one-parameter regression equation. Good correlations of GC retention data with first-order molecular connectivity indices were also reported [104,105].

Sporadically the reports appear on some advantages of Wiener number in QSRR over connectivity indices [106,107]. Opposite opinion is advocated in other publications [88,91]. High intercorrelations between the two types of the molecular graph-derived descriptors do not allow for unequivocal conclusions, however.

7.7. Simple Correlations of Retention with Substituent Constants

Reports on linear plots relating electronic substituent constants to GC retention parameters have been published since 1958 [108–116]. Regression analysis was first applied in 1978 by Nieudorp *et al.*, to describe the differences of logarithms of retention volume determined on *Apiezon L* and *Carbowax 2M* of a series of aromatic solutes to Hammett sigma values of substituents at three positions of substitution. Several other historically relevant references are listed elsewhere [37]. At least in some cases, the reliability of the approach may be questioned because there are too few substituent constant data points differing in magnitude to prove the significance of their changes for retention. Besides, different kinds of electronic substituent constants are added to one another or treated as fully exchangeable. For example, in a QSRR study of a series of substituted phenols the authors [118] added classical Hammett substituent constants for the meta and para positions (common practice) together with the Taft constants for the ortho position. The question arises as to what may be the physical meaning of an arithmetic sum of different substituent constants.

In normal phase TLC on silica gel correlations were found between R_M values and electronic constants of substituents in the phenyl ring of 5-arylidenbarbiturates [119,120]. The correlations were significant when the compounds studied were divided into subgroups by the type and position of a substituent. As a matter of fact, the predictive potency of the relationships is too low for analytical purposes but the regularities found can be helpful in discussing the mechanism of TLC adsorption.

Buydens *et al.* [91,121,122] studied relationships between the Hammett sigma and GC retention on nonpolar stationary phases. Important, however expected, conclusion is that the electronic substituent constant was a significant parameter in regression equations obtained. However, when the quantum chemical electronic parameters are included in the regression, they are preferred to sigma constants.

Electronic substituent constants were demonstrated to provide significant input to multiparameter QSRR, along with dispersive structural descriptors. These QSRR will be discussed in next chapter. It is difficult, however, to find appropriate chromatographic systems and large enough sets of solutes for which polar interactions would solely determine separations. It was much easier to isolate effects of solute bulkiness (dispersive interactions) on retention by selecting a nonpolar stationary phase and a set of solutes of the same polarity (*e.g.*, possessing the same functional group) but of varying nonpolar fragment. To overcome this difficulty an approach reported by Steurbaut *et al.* [123] may be considered. These authors analysed increments by individual substituents to GC and TLC retention parameters determined on various stationary phases for a group of *O*-alkyl-*O*-arylphenylphosphonothioates. These increments were related to Hammett constants. The increments, ΔR_M, determined by TLC poorly correlated to σ. The best was the correlation noted for ΔR_M obtained on polyamide with *n*-hexane-acetic acid 95:5 (v/v) as eluent ($r = 0.77$). Also in case of GC data, even if obtained on a highly polar column (DEGS), the correlations were low ($r = 0.76$).

Hansch π lipophilic substituent constant was correlated with ΔR_M from TLC already in 1965 [124]. The parameter ΔR_M quantifies the effect of a substituent upon the R_M value of a parent, unsubstituted compound. Latter on a correlation was also reported between the Zahradnik β constants (analogous to π constants) and R_M values [125]. Since early 1970s in RP TLC [126] and since mid 1970s in RP HPL [127], QSRR publications involving π parameter started to appear on regular basis. Early literature has been listed in ref. 37.

Carlson *et al.* [127] considered $\Delta \log k'$, *i.e.*, $\log k'$ of the substituted solute minus $\log k'$ of unsubstituted standard. They noted that correlations with π values for the respective substituents were good only if the two classes of solutes studied (*i.e.*, phenols and anilines) were analysed separately.

Kuchař *et al.* [128] obtained equations relating R_M values to π values with R_M data of some β-aryl-*n*-butyric acids derived from a chromatographic system where the acids would be non-ionized. In case of a system where the acids were ionized the correlation was poor but it improved after introducing σ constant as second regressor. Important finding by Kuchař *et al.* [129] was a significant departure from linearity in the relationship between π and R_M when the solutes studied represented the broader region of hydrophobicity. In another paper of the series Kuchař *et al.* [130] analysed RP HPLC data of their series of arylaliphatic acids. In RP HPTLC the chromatographic plates were either silanized or impregnated with silicone oil and buffer-acetone mixtures were the mobile phases. In RP HPLC, an octadecylsilica column was used with methanol-buffer eluent. The authors [130] claimed that extrapolation of retention parameters to pure water was not advantageous and negatively influenced the correlation between R_M and π.

There have been several publications at the early stage of development of QSRR methodology in which correlations between TLC and HPLC data on one hand, and π parameter on the other hand concerned homologous or closely related solutes. Obviously, equally good correlations could be obtained with carbon number and such publications are not disccused here. When the solutes differ structurally, the correlations between chromatographic data and π usually deteriorate. For example, the relationships between RP HPLC capacity factor and π for different phenyl-substituted phenyl carbamoylmethyliminodiacetic acids divide the solutes into three groups whose membership is determined by the degree of ortho substitution [131]. Similarly, separate QSRR had to be derived, describing the relationship between the π parameter and the contribution of a given substituent to the RP HPLC capacity factor, for C^2-, C^6- and C^8-substituted cyclic nucleotides [132]. The lack of linearity of $\log k'$ versus π plots have been reported for α-melanotropin analogues [133].

The hydrophobicity of a series of derivatives may be compared by summation of constants of the substituents present in a given multiple-substituted parent structure. Guerra *et al.* [134] reported a moderate correlation between RP HPLC capacity factors and $\sum\pi$ for a series of demorphin-related oligopeptides. On the other hand, an excellent correlation of $\log k'$ against $\sum\pi$ was found a series of phenylaliphatic acids [45]. To arrive at such a good correlation the authors [45] corrected their RP HPLC retention data for dissociation. In effect, the π values were related to $\log k'$ values for neutral molecules. This is a correct approach because the reference hydrophobicity parameters are valid only for non-dissociatied species. This problem will be discussed in details in Chapter 11 devoted to chromatographic determinations of hydrophobicity. In the same chapter the $\log P$ calculated from hydrophobic fragmental constants (*CLOGP*) will be analysed as a QSRR regressor.

The Hansch substituent hydrophobic constant π was demonstrated to be a valuable structural descriptor in multiparameter QSRR which will be discussed in Chapter 10.

7.8. Simple Correlations of Retention with Empirical Physicochemical Data

There are no reports on advantages in QSRR analysis of experimentally measured dipole moments and molar refractivities over the calculated ones [37].

A parameter of purely empirical nature tested occasionally in QSRR is boiling temperature. Correlation between GC specific retention volumes, V_g, and boiling point at 760 Torr, T_b, was studied by Castello and D'Amato [135]. The set of test solutes formed straight and branched bromoalkanes. The curvilinear relationship is apparent on the graphs presented by the authors [135], although the data points are scattered independent of the stationary phase on which the V_g data has been determined. The respective plots of $\ln V_g$ versus the number of carbon atoms in the subgroup of 1-bromoalkanes studied are precisely linear.

QSRR equations involving boiling points of solutes as regressors were reported in several publications by a group of Spanish researchers. In a relatively recent paper [136] good correlations are given between Kováts indices of a family of mono- and polycycloolefins possessing one or more endo- or exocyclic double bonds with their boiling points.

An interesting retention-physicochemical property relationship has recently been found by Pietrogrande *et al.* [137]. These authors got an apparently linear plot for 6 flavonoid compounds of the substituent contributions to RP HPLC retention, $\Delta \log k'$ versus ^{13}C NMR chemical shifts. According to the authors the relationship observed can result from differences in intermolecular hydrogen bonding properties among the solutes.

Mutual correlations between retention indices determined in different chromatographic systems simply reflect similarities in separation mechanism. QSRR relating two series of retention data determined on polar and nonpolar stationary phases will be discussed in Chapter 10.

There is one experimentally determined physicochemical parameter which has been extensively studied as a chromatographic retention predictor. This is the logarithm of *n*-octanol-water partition coefficient. Respective QSRR are exhaustively discussed in Chapter 11.

References

1. Jinno, K., "*A Computer-Assisted Chromatography System*", Huthing, Heidelberg, 1990.
2. Glajch, J.L. and Snyder, L.R. (Eds.), "*Computer-Assisted Method Development for High-Performance Liquid Chromatography*", Elsevier, Amsterdam, 1990.
3. Lochmüller, C.H., Reese, C., Aschman, A.J. and Breiner, S.J., *J. Chromatogr. A* **656**, 3 (1993).
4. Hamoir, T., De Smet, M., Pyrins, H., Conti, P., Vanden Driesche, N., Massart, D.L., Maris, F., Heindriks, H. and Schoenmakers, P.J., *J. Chromatogr.* **589**, 31 (1992).
5. Snyder, L.R., Carr, P.W. and Rutan, S.C., *J. Chromatogr. A* **656**, 537 (1993).
6. Martin, A.J.P., *Biochem. Soc. Symp.* **3**, 4 (1949).
7. Colin, H. and Guiochon, G., *J. Chromatogr. Sci.* **18**, 54 (1980).
8. Smith, R.M., *J. Chromatogr. A* **656**, 381 (1993).
9. Jandera, P., *J. Chromatogr. A* **656**, 437 (1993).
10. Gilpin, R.K., *J. Chromatogr. A* **656**, 217 (1993).

11. Carr, P.W., Li, J., Dallas, A.J., Eikens, D.J. and Tan, L.C., *J. Chromatogr. A* **656**, 113 (1993).
12. Prausnitz, J.M., "*Molecular Thermodynamics of Fluid Phase Equilibria*", Prentice Hall, Englewood Cliffs, 1986.
13. Park, J.H., Jang, M.D., Chae, J.J., Kim, H.C. and Suh, J.K., *J. Chromatogr. A* **656**, 69 (1993).
14. Tchapla, A., Colin, H. and Guiochon, G., *Anal. Chem.* **56**, 621 (1984).
15. Tchapla, A., Heron, S., Lesellier, E. and Colin, H., *J. Chromatogr. A* **656**, 81 (1993).
16. Ladon, A. and Sandler, S., *Anal. Chem.* **45**, 921 (1973).
17. Castello, G. and D'Amato, G., *J. Chromatogr.* **131**, 41 (1977).
18. Hoshikawa, Y., Koike, K. and Kuriyama, T., *Bunseki Kagaku* **21**, 307 (1972).
19. Smith, R.J., Haken, J.K. and Wainwright, M.S., *J. Chromatogr.* **334**, 95 (1985).
20. Budahegyi, M.V., Lombosi, E.R., Lombosi, T.S., Mèszarós, S.Y., Nyiredy, S., Tarjan, G., Timár, I. and Takacs, J.M., *J. Chromatogr.* **271**, 213 (1983).
21. Haken, J.K. and Korhonen, I.O.O., *J. Chromatogr.* **319**, 131 (1985).
22. Golovnya, R.U. and Grigoreva, D.N., *Zh. Anal. Khim.* **40**, 316 (1985).
23. Ellrén, O., Peetre, I.-B. and Smith, B.E.F., *J. Chromatogr.* **88**, 295 (1974).
24. Chernoplekova, V.A., Korol, A.N., Sokodynskii, K.I., Lopatina, U.S. and Kocheshkov, K.A., *Zh. Anal. Khim.* **30**, 1285 (1975).
25. James, A.T. and Martin, A.J.P., *Biochem. J.* **50**, 679 (1952).
26. Preisler, J., *Z. Anal. Chem.* **240**, 389 (1968).
27. Korol, A.N., *Uspekhy Khim.* **60**, 1225 (1982).
28. Wang, F., Jiang, Y., Jiang, D. and Wang, W., *J. Chromatogr. Sci.* **33**, 71 (1995).
29. Wasiak, W., *J. Chromatogr. A* **653**, 63 (1993).
30. Green, J., Marcinkiewicz, S. and McHale, D., *J. Chromatogr.* **10**, 158 (1963).
31. Clifford, D.R., Fieldgate, D.M. and Watkins, D.A.M., *J. Chromatogr.* **43**, 110 (1969).
32. Boyce, C.B.C. and Milborrow, B.V., *Nature* **208**, 537 (1965).
33. Murakami, F., *J. Chromatogr.* **198**, 241 (1980).
34. Vitali, T., Gaetani, E., Laureri, C.F. and Branca, C., *Il Farmaco (Ed. Sci.)* **31**, 58 (1976).
35. Koopmans, R.E. and Rekker, R.F., *J. Chromatogr.* **285**, 267 (1984).
36. Tomlinson, E., Poppe, H. and Kraak, J.C., *Int. J. Pharm.* **7**, 225 (1981).
37. Kaliszan, R., "*Quantitative Structure-Chromatographic Retention Relationships*", Wiley, New York, 1987.
38. Gudzinowicz, B.J. and Martin, H.F., *Anal. Chem.* **34**, 648 (1962).
39. Wurst, M., *Microchim. Acta* **1966**, 379 (1966).
40. Ellrén, O., Peetre, I.-B. and Smith, B.E.F., *J. Chromatogr.* **93**, 383 (1974).
41. Putnam, R.C. and Pu, H., *J. Gas Chromatogr.* **3**, 160 (1965).
42. Voroblev, L.N., *Coll. Czech. Chem. Commun.* **27**, 1045 (1962).
43. Araki, T., *Bull. Chem. Soc. Japan* **36**, 879 (1963).
44. Iwata, R., Makide, Y. and Tominaga, T., *Bull. Chem. Soc. Japan* **47**, 3071 (1974).
45. Macek, J. and Smolková-Kuelemansová, E., *J. Chromatogr.* **333**, 309 (1985).
46. Tomková, H., Kuchař, M., Rejholec, V., Pacákova, V. and Smolková-Kuelemansová, E., *J. Chromatogr.* **329**, 113 (1985).
47. Kuchař, M., Tomková, H., Rejholec, V. and Skalická, O., *J. Chromatogr.* **333**, 21 (1985).
48. Bortnikov, G.N., Egorochkin, A.N., Vyazankin, N.S., Chernyshev, E.A. and Yashin, Y.I., *Izv. Akad. Nauk SSSR Ser. Khim.* 1402 (1970).
49. Shatz, V.D., Sturkovich, R.Y. and Lukevics, E., *J. Chromatogr.* **165**, 257 (1979).
50. Lamparczyk, H., Wilczyńska, D. and Radecki, A., *Chromatographia* **17**, 301 (1983).
51. Lamparczyk, H. and Radecki, A., *Chromatographia* **18**, 615 (1984).
52. Wurst, M. and Churacek, J., *J. Chromatogr.* **70**, 1 (1972).
53. Saura Calixto, F. and Garcia Raso, A., *Chromatographia* **15**, 771 (1982).
54. Bermejo, J., Canga, J.S., Gayol, O.M. and Guillén. M.D., *J. Chromatogr. Sci.* **22**, 252 (1984).
55. Jinno, K. and Kawasaki, K., *Chromatographia* **17**, 337 (1983).
56. Jinno, K. and Kawasaki, K., *J. Chromatogr.* **316**, 1 (1984).
57. Jinno, K. and Kawasaki, K., *Chromatographia* **18**, 90 (1984).
58. Hanai, T., Jukurogi, A. and Huber, J., *Chromatographia* **19**, 266 (1984).
59. Bojarski, J. and Ekiert, L., "*Proceedings of the Symposium on Advanced Liquid Chromatography*", Akadèmiai Kiadó, Budapest, 1984.
60. James, A.T., *Anal. Chem.* **28**, 1564 (1956).
61. Ainshtein, A.A. and Shulyatieva, T.I., *Zh. Anal. Khim.* **27**, 816 (1972).
62. Shatz, V.D., Belikov, V.A. and Zelchan, G.I., *J. Chromatogr.* **174**, 83 (1979).
63. Siborov, R.I., Romanenko, L.C. and Reznikov, S.A., *Zh. Phiz. Khim.* **51**, 2913 (1977).
64. Anvaer, B.I., Zhukovitskii, A.A., Litovtseva, I.I., Sakharov, V.M. and Turkeltaub, N.M., *Zh. Anal. Khim.* **19**, 178 (1964).

65. Abegaz, B. and Gerba, S., *J. Chromatogr.* **324**, 440 (1985).
66. Ong, V.S. and Hites, R.A., *Anal. Chem.* **63**, 2829 (1991).
67. Scott, R.P.W., *J. Chromatogr.* **122**, 35 (1976).
68. Karger, B.L., Snyder, L.R. and Eon, C., *J. Chromatogr.* **125**, 71 (1976).
69. Janini, G.M., Johnston, K. and Zieliński, W.L. Jr., *Anal. Chem.* **47**, 670 (1975).
70. Zieliński W.L. and Janini, G.M., *J. Chromatogr.* **186**, 237 (1979).
71. Kaliszan, R., Lamparczyk, H. and Radecki, A., *Biochem. Pharmacol.* **28**, 123 (1979).
72. Wise, S.A., Bonnett, W.J., Guenther, F.R. and May, W.E., *J. Chromatogr. Sci.* **19**, 457 (1981).
73. Sander, L.C. and Wise, S.A., *Crit. Rev. Anal. Chem.* **18**, 299 (1987).
74. Randić, M., *J. Am. Chem. Soc.* **97**, 6609 (1975).
75. Karger, B.L., Gant, J.R., Harthopf, A. and Weiner, P.H., *J. Chromatogr.* **128**, 65 (1976).
76. Kaliszan, R. and Foks, H., *Chromatographia* **10**, 346 (1977).
77. Kaliszan, R., *Chromatographia* **10**, 529 (1977).
78. Heintz, M., Gruselle, M., Druilhe, A. and Lefort, D., *Chromatographia* **9**, 367 (1976).
79. Michotte, Y. and Massart, D.L., *J. Pharm. Sci.* **66**, 1630 (1977).
80. Kaliszan, R. and Lamparczyk, H., *J. Chromatogr. Sci.* **16**, 246 (1978).
81. Millership, J.S. and Woolfson, A.D., *J. Pharm. Pharmacol.* **30**, 483 (1978).
82. Randić, M., *J. Chromatogr.* **161**, 1 (1978).
83. Bonchev, D., Mekenjan, O., Protić, G. and Trinajstić, N., *J. Chromatogr.* **176**, 149 (1979).
84. Whalen-Pedersen, E.K. and Jurs, P.C., *Anal. Chem.* **53**, 2184 (1981).
85. Bartle, K.D., Lee, M.L. and Wise, S.A., *Chromatographia* **14**, 69 (1981).
86. Vassilaros, D.L., Kong, R.C., Later, D.W. and Lee, M.L., *J. Chromatogr.* **252**, 1 (1982).
87. Kier, L.B. and Hall, L.H., *J. Pharm. Sci.* **68**, 120 (1979).
88. Sabljić, A., *J. Chromatogr.* **319**, 1 (1985).
89. Kaliszan, R., *Chromatographia* **12**, 171 (1979).
90. Huang, J.-X., Bouvier, E.S.P., Stuart, J.D., Melander, W.R. and Horváth, C., *J. Chromatogr.* **330**, 181 (1985).
91. Buydens, L. and Massart, D.L., *Anal. Chem.* **53**, 1990 (1981).
92. Wells, M.J.M., Clark, C.R. and Patterson, R.M., *J. Chromatogr. Sci.* **19**, 573 (1981).
93. Stead, A.H., Gill, R., Evans, A.T. and Moffat, A.C., *J. Chromatogr.* **234**, 277 (1982).
94. Bojarski, J. and Ekiert, L., *J. Liq. Chromatogr.* **6**, 73 (1983).
95. Doherty, P.J., Hoes, R.M., Robbat, A., Jr. and White, C.M., *Anal. Chem.* **56**, 2697 (1984).
96. Raymer, J., Wiesler, D. and Novotny, M., *J. Chromatogr.* **325**, 13 (1985).
97. Bojarski, J. and Ekiert, L., *Chromatographia* **15**, 172 (1982).
98. De Voogt, P. and Govers, H., *Chemosphere* **15**, 1467 (1986).
99. Hale, M.D., Hileman, F.D., Mazer, T., Shell, T.L., Noble, R.W. and Brooks, J.J., *Anal. Chem.* **57**, 640 (1985).
100. Dunn, W.J., Koehler, M., Stalling, D.L. and Schwartz, T.R., *Anal. Chem.* **58**, 1835 (1986).
101. Ośmiałowski, K. and Kaliszan, R., *Quant. Struct.-Act. Relat.* **10**, 125 (1991).
102. Kupchik, E.J., *J. Chromatogr.* **630**, 223 (1993).
103. Kupchik, E.J., *J. Chromatogr.* **645**, 182 (1993).
104. Dmitrikov, V.P. and Nabivach, V.M., *Zh. Anal. Khim.* **42**, 1110 (1987).
105. Buryan, P., Nabivach, V.M. and Dmitrikov, V.P., *J. Chromatogr.* **509**, 3 (1990).
106. Bošnjak, N., Mihalić, Z. and Trinajstić, N., *J. Chromatogr.* **540**, 430 (1991).
107. Amić, D., Davidović-Amić, D. and Trinajstić, N., *J. Chromatogr. A* **653**, 115 (1993).
108. Jones, W.L. and Kieselbach, R., *Anal. Chem.* **30**, 1590 (1958).
109. Langer, S.H., Zahn, C. and Pantazoplos, J., *J. Chromatogr.* **3**, 154 (1960).
110. Gaile, A.A. and Leitman, Y.I., *Zh. Phiz. Khim.* **41**, 2886 (1967).
111. Martire, D.E. and Riedl, P., *J. Phys. Chem.* **72**, 3478 (1968).
112. Karger, B.L., Elmehrik, I. and Stern, R.L., *Anal. Chem.* **40**, 1227 (1968).
113. Stern, R.L., Karger, B.L., Keane, W.J. and Rose, H.C., *J. Chromatogr.* **39**, 17 (1969).
114. Ono, A., *Nippon Kagaku Zasshi* **92**, 429 (1971).
115. Parizheva, I.V., Gaile, A.A. and Prokuryakov, V.A., *Zh. Phiz. Khim.* **48**, 2533 (1974).
116. Zakharov, A.P., Gaile, A.A. and Prokuryakov, V.A., *Zh. Phiz. Khim.* **51**, 2108 (1977).
117. Nieudorp, G.H.E., de Ligny, C.L. and van der Keen, N.G., *J. Chromatogr.* **154**, 133 (1978).
118. Radecki, A., Grzybowski, J., Lamparczyk, H. and Nasal, A., *J. High Resolut. Chromatogr. Chromatogr. Commun.* **2**, 581 (1979).
119. Mokrosz, J.L. and Ekiert, L., *Chromatographia* **18**, 401 (1984).
120. Ekiert, L., Bojarski, J. and Mokrosz, J., in: "*Chromatography, the State of the Art*", Kalász H. and Ettre L.S. (Eds.), Akadémiai Kiadó, Budapest, 1985.
121. Buydens, L., Massart, D.L. and Geerlings, P., *Anal. Chem.* **55**, 738 (1983).

122. Buydens, L., Coomans, D., Vanbelle, M., Massart, D.L. and Vanden Driesche, R., *J. Pharm. Sci.* **72**, 1327 (1983).
123. Steurbaut, W., Dejonckhere, W. and Kips, R.H., *J. Chromatogr.* **160**, 37 (1978).
124. Iwasa, J., Fujita, T. and Hansch, C., *J. Med. Chem.* **8**, 150 (1965).
125. Kopecky, J. and Boček, K., *Experientia* **23**, 125 (1967).
126. Biagi, G.L., Guerra, M.C. and Barbaro, A.M., *J. Med. Chem.* **13**, 944 (1970).
127. Carlson, R.M., Carlson, R.E. and Kopperman, H.L., *J. Chromatogr.* **107**, 219 (1975).
128. Kuchař, M., Brůnová, B., Rejholec, V. and Rábek, V., *J. Chromatogr.* **92**, 381 (1974).
129. Kuchař, M., Rejholec, V., Jelinkova, M., Rábek, V. and Nemecek, O., *J. Chromatogr.* **162**, 197 (1979).
130. Kuchař, M., Rejholec, V., Kraus, E., Miller, V. and Rábek, V., *J. Chromatogr.* **280**, 279 (1983).
131. Nunn, A., *J. Chromatogr.* **255**, 91 (1983).
132. Braumann, T. and Jastorff, B., *J. Chromatogr.* **350**, 105 (1985).
133. Cody, W.L., Wilkes, B.C. and Hruby, V.J., *J. Chromatogr.* **314**, 313 (1984).
134. Guerra, M.C., Barbaro, A.M., Cantelli Forti, G., Pietrogrande, M.C., Borea, P.A. and Biagi, G.L., *J. Liq. Chromatogr.* **7**, 1495 (1984).
135. Castello, G. and D'Amato, G., *J. Chromatogr.* **324**, 363 (1985).
136. Diez, M.A., Guillén, M.D., Blanco, C.G. and Bermejo, J., *J. Chromatogr.* **508**, 363 (1990).
137. Pietrogrande, M.C., Reschiglian, P., Dondi, F., Kahie, Y.D. and Bertolasi, V., *J. Chromatogr.* **592**, 65 (1992).

8 Multiparameter QSRR Involving Empirical and Semiempirical Solute Descriptors

8.1. Predictive QSRR

In this chapter those selected QSRR are discussed which were derived in order to predict retention of structurally defined solutes at given chromatographic conditions. As a matter of fact, vast majority of such QSRR have questionable practical predictive value. They were derived either retrospectively for a set of published retention data or for a series of solutes which at the same time served to check the reliability of the predictions. Not many published QSRR were used to solve specific analytical tasks, like identification of an unknown peaks on a chromatogram. Thus, the so-called predictive QSRR are rather important from the point of view of detecting of structural features (or descriptors) determining differences in retention of solutes.

Predictive QSRR have been of unquestionable value for improving of chemometrics methodology and experimental skills. Certainly, some reliable predictive QSRR provide information on mechanism of separation operating at given chromatographic conditions. These QSRR will be discussed in Chapter 10. The problem is however, in determining which aspect of reported QSRR is prevailing. The judgement is to a large extent subjective. Subjective is also the selection of the most representative QSRR among multitude of published ones.

The following text is to illustrate achievements and limitations of predictive QSRR. Even if one agrees with Evans and Haken [1,2] that despite much work and many reports it is obvious that no realistic scheme of wide application is available for precalculation of retention indices, the informative value of QSRR for chemistry remains undisputable.

8.2. Substituent Constants as Retention Descriptors

Multiparameter QSRR involving the Hammett constant, the Taft constant and the resonance constant were proposed in 1975 to predict GC retention [3]. A combination of electronic, steric and hydrophobic substituent constants to describe GC retention indices was also tested by Haken *et al.* [4]. These early QSRR may be questioned with regard to statistical significance of individual terms in the regression equations derived.

Jinno and Kawasaki [5] attempted to describe the $\log k'$ from RP HPLC for a series of substituted phenols by means of the following equation in terms of the Hammett σ and Hansch π:

$$\log k' = 0.177\pi + 0.182\pi\sigma(1-\pi) - 0.367. \tag{8.1}$$

Equation (8.1) appears to be "predictive" (multiple correlation coefficient $R = 0.954$). The problem is, however, what physical meaning could be assigned to the product of π and σ. Formally correct is the relationship given by Yamagami *et al.* [6] describing RP HPLC capacity factors in terms of π and δ for a group of meta- and para-substituted phenylacetanilides. Zhou *et al.* [7] reported QSRR describing $\ln k'_w$ from RP HPLC in terms of substituent constants: Hansch π, Hammett σ, Taft E_s and Swain-Lupton F. Unfortunately, no information on the significance of individual terms in multiple regression equation was provided.

Of limited statistical value is also a recent study by Kim *et al.* [8] in which the Swain-Lupton field and resonance constants are the regressors along with a parameter called by the author the nonpolar bonding constant. The reason is that only five or eight RP HPLC data were used to derive QSRR.

8.3. Functional Group Contributions to Retention

In 1950 Martin [9] suggested that substituents change retention of a parent compound by factors which depend on the nature of the substituents and the separation system but are independent of the parent structure. First systematic studies on substituent increments to retention were the subject of several reports by Green *et al.* [10].

Martin rule can be expressed [11] by the equation,

$$\log k'_S = \log k'_P + \sum_{i=1}^{n} \tau_i, \tag{8.2}$$

where k'_P is the retention parameter of a parent compound, k'_S is the corresponding value for the derivative carrying n substituents and τ are retention increments due to individual substituents i.

Among the numerous reports concerning fragmental contributions to GC retention indices, the majority deal with hydrocarbons as a group of solutes and CH_2 as the fragment. They will not be discussed here. The report [11] concerning polychlorinated biphenyls should perhaps be mentioned although these solutes resemble hydrocarbons with respect to their chromatographic behaviour.

An approach described by Tejedor [12] deserves interest as it allows the calculation of GC retention indices of mono- and dialkylaromatic compounds on stationary phases whose polarity lies between that of *Squalane* and that of *Ucon SOHB* 280*X* (and that corresponds to Rohrschneider constants of between 0 and 1.71). Tejedor [12] calculated the contributions to retention of normal and branched alkyl groups as well as the dependence of these contributions on the polarity of the stationary phase. However, attempts to apply the regularities found for prediction of retention of the hydrocarbons studied on more polar stationary phases (*Apiezon L*, *Carbowax* 20*M*) were unsuccessful.

A group of Hungarian researchers worked for many years on a method for a *priori* calculations of GC retention indices, allowing the assessment of the contributions of atoms and bonds and interactions with stationary phases. The results of numerous publications by the group have been summarized in 1983 [13]. The basic assumption has been that retention index can be divided into three components

$$I^{\text{St.Ph.}}{}_{\text{Substance}}(T) = I_a + I_b + I^{\text{St.Ph.}}{}_{\text{Int}}(T), \tag{8.3}$$

where $I^{\text{St.Ph.}}{}_{\text{Substance}}$ is the retention index (in retention index units) determined under isothermal conditions for a given substance on a given stationary phase; T is the column temperature (°C or K); I_a is the atomic index contribution; I_b is the bond index contribution and $I^{\text{St.Ph.}}{}_{\text{Int}}$ is the interaction index contribution.

To calculate the bond increments the energy of individual bond is multiplied by the factor 0.018 index unit per kJ. For example,

$$\begin{aligned} I_b(\text{Ethane}) &= I_b(\text{C–C}) + 6I_b(\text{C–H}) = 349.1\,\text{kJ} \times 0.018\,\text{i.u./kJ} \\ &\quad + 412.8\,\text{kJ} \times 0.018\,\text{i.u./kJ} = 50.86\,\text{i.u.} \end{aligned} \tag{8.4}$$

The atomic index increment I_a is

$$I_a = \sum_{g=1}^{h} (\text{Atomic Mass}/10)_g = \text{Molecular Mass}/10, \tag{8.5}$$

where g is a serial number and h is the number of atoms in the molecule. For example,

$$I_a(\text{Ethane}) = (6 \times 1 + 2 \times 12)/10 = 3.00\,\text{i.u.} \tag{8.6}$$

The interaction index contribution to Eq. (8.3) is composed of two parts

$$I^{\text{St.Ph.}}{}_{\text{Int}}(T) = I^{\text{St.Ph.}}{}_{i,g}(T) + I^{\text{St.Ph.}}{}_{i,i}(T), \tag{8.7}$$

where $I^{\text{St.Ph.}}{}_{i,g}(T)$ is the general interaction index contribution and $I^{\text{St.Ph.}}{}_{i,i}(T)$ is the individual interaction index contribution.

The general interaction index contribution is calculated at a given column temperature T from the molecular index contribution, $I_M = I_a + I_b$, and the solute vapour pressure index, $I^0(s, T)$:

$$I_{i,g}(s, T) = I^0(s, T) - I_a - I_b. \tag{8.8}$$

The vapour pressure index is calculated from the formula

$$I^0(s,T) = 100\{[\log p^0(s) - \log p^0(z)]/[\log p^0(z+1) - \log p^0(z)] + z\}, \quad (8.9)$$

where z and $(z+1)$ mean carbon number in the reference homologues. The general interaction index contribution is obtained by the following summation:

$$I_{i,g}(s,T) = \sum I_{i,g}(g)_T, \quad (8.10)$$

where g is a serial number and q is the number of increments in the molecule. For a given solute s and the stationary phase *St.Ph.* the individual interaction index contribution is

$$I^{\mathrm{St.Ph.}}{}_{i,i}(s,T) = \sum_{g=1}^{q} I_{i,i}(g)^{\mathrm{St.Ph.}}{}_T, \quad (8.11)$$

or

$$I^{\mathrm{St.Ph.}}{}_{i,i}(s,T) = I^{\mathrm{St.Ph.}}(s,T) - I^0(s,T). \quad (8.12)$$

The retention indices predicted by the above described method from the structure of solute molecules are often in good agreement with experimental values. Unfortunately, the applicability of the method is limited to a very narrow range of compounds like isoalkanes.

GC retention index increments for substituent groups of multiple-substituted pyrazine derivatives have been determined by Mihara and Enomoto [14]. The increments were assigned separately for the *Carbowax* 20*M* and the OV-101 phases.

Osiek *et al.* [15] attempted to predict GC retention indices of a series of amidines on an SE-30 nonpolar column. These authors reported the prediction of Kováts indices based on regression equations involving the corresponding retention indices of two structural elements constituting the solute molecule.

Chen and Horvath [16] presented a general approach to the evaluation of substituent contributions to chromatographic retention by using a multiple linear regression analysis and indicator variables. Retention data, κ_i ($\log k'$ from HPLC or R_M from TLC) of n congeners containing m substituents were expressed by a set of linear equations:

$$\kappa_1 = \kappa_P + I_{11}\cdot\tau_1 + I_{12}\cdot\tau_2 + \cdots + I_{1m}\cdot\tau_m,$$

$$\kappa_2 = \kappa_P + I_{21}\cdot\tau_1 + I_{22}\cdot\tau_2 + \cdots + I_{2m}\cdot\tau_m,$$

$$\ldots \quad \ldots \quad \ldots \quad \ldots \quad \ldots \quad \ldots \quad \ldots \quad \ldots$$

$$\ldots \quad \ldots \quad \ldots \quad \ldots \quad \ldots \quad \ldots \quad \ldots \quad \ldots$$

$$\kappa_m = \kappa_P + I_{n1}\cdot\tau_1 + I_{n2}\cdot\tau_2 + \cdots + I_{nm}\cdot\tau_m, \quad (8.13)$$

where τ_{ji} is a substituent parameter which measures the change in chromatographic retention upon replacing a hydrogen atom in a substance i by substituent j. The maximum number of substituent parameters is m for the set of congeners. The τ_{ij} values for a given substituent in a certain position are the same in different compounds i, *i.e.*, $\tau_{ij} = \tau_j$. A given substituent in different positions, however, may have different τ values, depending on the particular molecular structure, and therefore the number of τ_j may be greater than the number of actual substituents. In the set of equations (Eq. (8.13)) κ_P is the retention value for the parent (non-substituted) compound. The coefficients I_{ij} are the indicator variables, which are set to unity when a τ_j value is assigned to a substituent in compound i, and are set to zero otherwise. For the parent compound all indicator variables I_{ij} in Eq. (8.13) are zero and for the congeners $I_{ij} = 0$ in case of a position without substituent. The matrix (Eq. (8.13)) is solved by using a multiple regression analysis.

In their publication Chen and Horváth [16] analysed the paper chromatography retention data obtained by Kuchař *et al.* [17] for aromatic-aliphatic acids. The R_M values predicted on the basis of the statistically evaluated ΔR_M values of substituents have been consistent with the observed data. Similar analysis was done with a set of aromatic amines, acids and amino acids chromatographed in RP HPLC systems. In this instance, the number of retention increments τ_j was greater than that of the substituents. The authors [16] have also found that the τ_j values calculated from the corresponding κ_i data obtained with three slightly different systems (octadecylsilica stationary phases: *Partisil, Spherisorb and LiChrosorb*; temperature 296 K and 343 K) show satisfactory correlation. Thus the prediction of retention data from substituent contributions appears possible for different chromatographic conditions.

Similarly to substituent contributions to retention the contributions of other structural features have been quantified. DiBussolo and Nes [18] determined such contributions to the HPLC retention for several dozen sterols with regard to cholesterol as the parent structure. The features examined by these authors included the presence and absence of methyl and ethyl groups, double bonds and hydroxyl groups in the nucleus and side chain, the chirality at C_3 and C_{24}, the configuration about certain double bonds, and the length of the side chain and branching. The comparison of the observed retention data for a given sterol with those calculated by means of structural contributions can be used to gain structural information on a sterol from its chromatographic behaviour.

The contributions of the functional group X to HPLC retention are usually defined as

$$\tau_X = \log k'_{R\text{-}X} - \log k'_{R\text{-}H} = \log(k'_{R\text{-}H}/k'_{R\text{-}H}). \qquad (8.14)$$

Riley *et al.* [19] found the τ values from RP HPLC to correlate with Hansch π constants. Another early finding by Tomlinson *et al.* [20] was that the τ values for different groups at different positions of substitution depended linearly on the mobile phase composition as expressed by the mobile phase surface tension. This relationship is in agreement with the solvophobic theory of chromatography [21]. Thus, the hypothetical τ values corresponding to 100% water mobile phase could be derived. These parameters are often reported and are usually denoted τ_w.

Morishita *et al.* [22] and Smith and Burr [23] proposed substituent indices $I_{S,X}$ determined from the difference in the HPLC retention indices of analytes on the addition of a substituent X,

$$I_{S,X} = 100(\tau_X/\tau_{CH_2}). \tag{8.15}$$

Methylene group increments were studied by several groups of researchers. It appeared that small but significant differences in the values of methylene increment is observed for individual series of homologues [24,25]. The methylene contributions corresponding to 100% water eluent, τ_w, vary depending on the nature of organic modifier in the eluents providing the polycratic retention data [26–28]. Based on a long series of experiments in strictly controlled conditions and using a single batch of stationary phase Smith and co-workers determined functional group contributions for a selection of aromatic substituents. It appeared that for eluent compositions, which give comparable capacity factors, there are significant differences of the contribution of some groups. For example the τ values of cyano, nitro, acetyl and carboxymethyl groups are -0.37, -0.17, -0.31 and -0.09, respectively, in 70% methanol; -0.27, -0.15, -0.31 and -0.09 in 50% acetonitrile but -0.42, -0.15, -0.53 and -0.24 in 30% tetrahydrofuran [29].

If there is more than one functional group in a solute molecule then these groups often interact electronically or by hydrogen bonding. In such a situation the τ values will represent a summed value representing the inherent group functional contribution and interaction contributions. Analysing HPLC data of protonated catecholamines Molnár and Horváth [30] derived τ values for the 3-OH group (-0.53 or -0.24 with 4-OH and 4-OH, respectively), 4-OH group (0.60 or -0.80, if no ionised amino group was present), 3-OCH_3 (0.52) and α-OH group (-0.94 or -0.58 depending on the absence or presence of an ionised amino group), α-NH_3^+ group (-0.89), β-NH_2^+-CH_3 (0.29) and β-CH_2 (0.29). A series of substituted (R) phenylethylamines Ph-βCH_2-αCH_2-NR_2 was studied by Gill *et al.* [31] using a methanol-buffer pH 3.15 (10:90 v/v) eluent. It was found that the addition of methyl had a larger effect in the α-position ($\tau = 0.36-0.44$) than on nitrogen ($\tau = 0.02-0.13$). Other authors also reported on the effects of position of substitution on HPLC retention [32–40]. In their studies on retention prediction of complex molecules Valko *et al.* [41] ignored the effects of different eluents and used average values of functional group contributions. Thus, for C-hydroxylation, N-demethylation and decarboxylation they assumed τ values of -0.313, -0.162 and -0.622, respectively.

To account for mutual changes in τ values caused by the presence in solute molecule of two or more functional groups some correction factors have been proposed. Tsantili-Kakoulidou *et al.* [42] related the differences due to substituent interactions to the substituent X and Y interaction parameters ρ and τ proposed by Brändstrom [43] and Fujita [44]:

$$T_I = \rho_x \sigma_y + \rho_y \sigma_x, \tag{8.16}$$

where T_I was the difference between the experimental $\log k'_w$ values from HPLC analysis of a solute and those obtained by the addition of substituent $\log k'_w$ values.

The authors [42] found the σ/ρ correction values to be of limited value for the description of the RP HPLC retention parameters of substituted phenols and anilines.

A similar attempt to quantify substituent interactions was undertaken by Smith and Burr [45]. These authors described the RP HPLC retention index I of disubstituted (X, Y) aromatic solutes employing the following equation:

$$I = I_P + I_{S,R} + \sum I_{S,\text{Al-}X} + \sum I_{S,\text{Ar-}X} + \sum I_{I,X\text{-}Y}, \tag{8.17}$$

where I_P represents the retention parameter of a parent non-substituted compound, $I_{S,R}$ is a contribution for saturated alkyl chains, $I_{S,\text{Al-}X}$ are contributions for substituents on saturated aliphatic carbons, $I_{S,\text{Ar-}X}$ are contributions for aromatic substituents and $I_{I,X\text{-}Y}$ are terms accounting for any interactions between substituents caused by electronic, hydrogen bonding and steric effects. The interaction terms are calculated by the following equation:

$$I_{I,X\text{-}Y} = (\sigma_x \rho_y^* + \sigma_y \rho_x^*) + F_{\text{HB}}^* + F_0^*, \tag{8.18}$$

where ρ^*, F_{HB}^* and F_0^* are expressed in units of the retention parameter; ρ^* are the susceptibilities of X and Y to the modifying effects of Y and X on the Hammett σ constants of the substituents σ_x and σ_y; F_{HB}^* is a term accounting for hydrogen bonding and F_0^* is a term reflecting the ortho effect. Encouraging correlations of the observed and predicted by Eq. (8.17). Retention parameters determined with eluents containing methanol deteriorated in case of the tetrahydrofuran-containing mobile phases [29].

Effects of substituent interactions on HPLC retention were determined by Spanjer and De Ligny [46,47]. The data obtained correlated with Hammett and Taft type functions. In GC, retention indices of substituted pyrazines were calculated from substituent contributions by Mihara and Masuda [48].

Numerous studies prove that the magnitude of functional group contributions to RP HPLC or RP TLC retention usually depend on the nature and percentage of the organic modifier in the eluent. Representative may be the plots of τ *versus* % methanol in a binary aqueous mobile phase taken from Smith [29] (Fig. 8.1).

Similar differences can also be observed in data reported by Tanaka *et al.* [49]. Huang *et al.* [32] and Dondi *et al.* [37,38]. This problem is important from the point of view of selectivity and optimization of separations. In QSRR often the functional group contributions are considered which correspond to 100% water eluent: the τ_w values. Obviously, when the τ *versus* % composition relationships are curved, the extrapolated τ_w value would be dependent on the range of composition selected for the calculation. Wells and Clark [50] found differences in τ_w for some functional groups depending on the nature of organic modifier. Still, however, such extrapolated standardised data appear most appropriate for comparisons with other parameters and properties.

Systematic changes in τ values with temperature of eluent were reported [19,51]. Different τ values for some functional groups were determined on different octadecyl-silica-bonded stationary phases [16,52]. Naturally, differences were also observed of

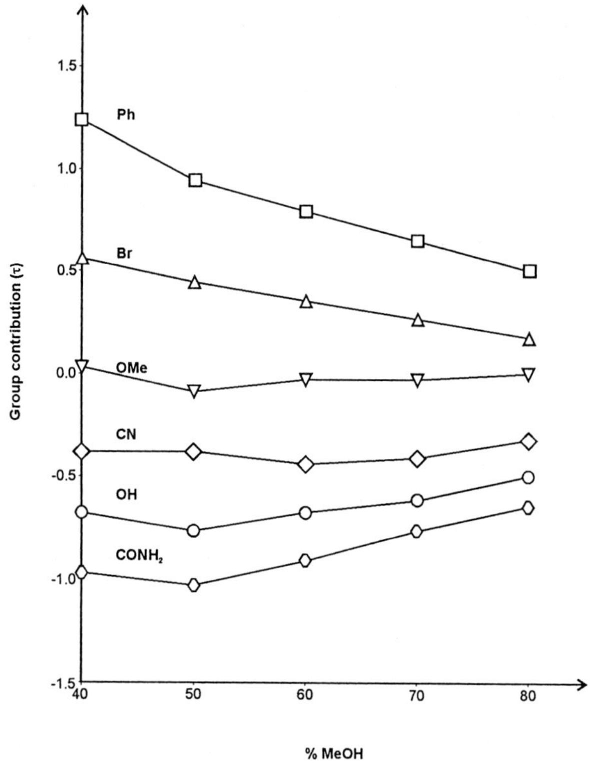

Figure 8.1. Effect of proportion of methanol in the water-methanol eluent on functional group contributions. (Reprinted from Smith, R.M., *J. Chromatogr. A* 656, 381, 1993 with kind permission of Elsevier Science-NL, Sara Burgerhartstraat 25, 1055 KV Amsterdam, The Netherlands.)

τ values determined on different reversed-phase columns, *e.g.*, polystyrene-divinylbenzene with regards to octadecylsilica [53,54].

The functional group contributions to retention, especially the extrapolated to pure water τ_w values from RP HPLC, were demonstrated to correlate well with other substituent constants, especially with Hansch π values [19,20,36,50,55,56].

In conclusion it can be reported after Smith [29] that the τ values can be used to predict the retention of simple substituted compounds based on parent structure. In case of polyfunctional compounds the interactions between substituents make such predictions of a rather limited value.

It has been assumed that the chromatographic distribution coefficient could be calculated from known values of the activity coefficients of the substance of interest in both chromatographic phases. A means for the assessment of activity coefficients is the *UNIFAC* group contribution method according to Fredenslund *et al.* [57,58].

The UNIFAC (UNIQUAC functional group activity coefficient) combines the UNIQUAC model of solutions [59] and the analytical solution of group (ASOG) concept [60]. The UNIQUAC (universal quasi-chemical) is an approximate model of liquid mixtures based on the quasi-chemical lattice model of Guggenheim [61]. The UNIFAC method transforms a solution of molecules into a solution of groups.

Activity coefficients are computed by virtue of structural parameters and binary group interaction parameters reflecting the energy of mutual interaction of functional groups present in the mixture. The structural parameters are calculated from the Van der Waals volumes and surface areas of functional groups [59]. The interaction parameters are derived from collections of experimental phase equilibrium data [62].

There have been several reports on application of UNIFAC to predicting GC and HPLC retention [63–72]. The approach was demonstrated to be useful for the prediction of relative retention, the relative strengths of solvents, the effects of the carbon number and of the functional groups of homologous series of solutes as well as other important chromatographic properties. However in general, the quantitative predictions of retention by UNIFAC are still not accurate enough to be useful [71,72].

A semi-empirical description of RP HPLC systems, allowing for predicting the relative retention and selectivity within a series of compounds, has been developed by Jandera. A suitable set of standard reference solutes is necessary to calibrate the retention (or selectivity) scale. The approach consists on determining the interaction indices and the structural lipophilic and polar indices. The theoretical background of the retention model was published in 1982 [73]. A subsequent series of papers by Jandera [74–77] established the approach.

The starting relationship in Jandera's RP HPLC model is

$$\log k' = a - mX + dX^2, \tag{8.19}$$

where X is the concentration of the organic modifier in a binary aqueous solvent; a, m and d are constants. The value of the quadratic term dX^2 in Eq. (8.19) often is not very significant and can be neglected to first approximation at least for lower values of X. The retention equation is then reduced to Eq. (8.20):

$$\log k' = a - mX. \tag{8.20}$$

To calibrate the retention scale, Jandera [74] measured the k' data of a homologous series of solutes at several concentrations of the organic solvent in the eluent. From the experimental data the constants a and m in Eq. (8.20) were calculated by regression analysis. Next, from the constants a and m, the other constants, a_0, a_1, q and p were calculated assuming the following relationships:

$$a = a_0 + a_1 n_c, \tag{8.21}$$

$$m = q + pa, \tag{8.22}$$

where n_c is the number of carbon atoms in the saturated aliphatic straight chain of calibration homologues. The constants in Eqs. (8.21) and (8.22) were calculated by linear regression using the a and m data for individual solutes of the calibration series.

Combining Eqs. (8.21) and (8.22) with Eq. (8.20) Jandera [74] got the following equation:

$$\log k' = (a_0 + a_1 n_c) - (q + pa)X = (a_0 + a_1 n_c)(1 - pX) - qX. \tag{8.23}$$

In systems where the quadratic form of $\log k'$ *versus* X relationships should be considered Jandera [74] modified his Eq. (8.23) into the form:

$$\log k' = (a_0 + a_1 n_c)(1 - pX) - qX + (d_0 + d_1 n_c)X^2, \tag{8.23}$$

where d_0 and d_1 are constants which may also be determined from the capacity factors of calibration homologues.

According to Jandera [75], a_1 and p do not depend significantly on the nature of the homologous series but on the organic modifier used as the less polar component of the mobile phase. On the other hand, a_0 and q depend also on the type of homologous series and on the stationary phase used. The constants a_0, a_1 and p, determined from k' data of calibration homologues, are considered to be generally valid also for all the other compounds. If this assumption is correct, then by Eq. (8.23) or (8.24), n_{ce} and q_i for an arbitrary compound may be calculated from its experimental k' measured at least at two different mobile phase compositions, X. Such determined constants n_{ce} and q_i are equivalent to the n_c and q of the calibration homologues and may be related to the lipophilic and polar contributions to retention, respectively.

Jandera [75–77] has practically verified his theoretical approach considering selectivity, α, in various RP HPLC systems. Experiments were carried out using several C_8 and C_{18} columns and binary solvents as well as ternary mobile phases. In order to allow a comparison of selectivity for various compounds under changing conditions, the retention of all compounds was related to a single standard compound with the capacity factor k'_S. Thus, for a given chromatographic system

$$\log \alpha = \log k'_i - \log k'_S \tag{8.25}$$

Using Eq. (8.23) or (8.24) for capacity factors in Eq. (8.25), one gets:

$$\log \alpha = a_1(1 - pX)\Delta n_c - X\Delta q, \tag{8.26}$$

or

$$\log \alpha = [a_1(1 - pX) + d_1 X^2]\Delta n_c - X\Delta q, \tag{8.27}$$

where Δ denotes the difference of the respective constants for a given solute and the standard compound. Jandera [75] assumes the first term in Eq. (8.26) or (8.27) to be the lipophilic contribution to selectivity, $\log \alpha_L$, which is controlled by the size of the nonpolar (hydrocarbon) part of the molecule and is proportional to Δn_c. The present author thinks it more appropriate to treat this term as depending on ability of the solute to participate in nonspecific dispersive interactions with components of a chromatographic system. The second term $(-X\Delta q)$ in Eqs. (8.26) and (8.27) is the polar contribution to selectivity, $\log \alpha_P$, which is proportional to the difference between the constants q_i of the solute and of the standard compound and to the concentration of the organic solvent in the mobile phase, X.

The Δn_c of a given test compound is the equivalent to the number of carbon atoms in the alkyl chain of the calibration homologues; $\log \alpha_L$ depends also on the character of both the column and the organic solvent used in the mobile phase. The value of log α_L decreases with the increasing concentration of the organic solvent in the mobile phase, X. The Δq characterizes the difference between the polarities of the functional groups in the molecules of the solute and of the standard. The value of log α_P can be attributed to the interactions of the chromatographed compounds with the organic solvent in the eluent.

Jandera reports [75–77] a good correlation ($r = 0.991$) of the indices n_{ce} or Δn_c with the Hansch hydrophobic substituent constant [78]. Similarly good ($r = 0.988$) is the correlation of the relative indices Δq with the Snyder polarity indices P' [79].

The relative polar indices Δq were found to increase with the polarities of the functional groups approximately in the order; n-alkanes < polycyclic aromatic hydrocarbons < n-alkylbenzenes < benzene, styrene, biphenyl < halogenated benzenes < dialkyl ethers < alkylaryl ethers, diaryl ethers < aromatic nitriles < aromatic ketones and aldehydes < aromatic amines < aromatic alcohols < phenols, alkylphenols < chlorophenols [77]. For a given class of compounds, the Δq values were found within a relatively narrow range [75].

The experimentally derived constants Δn_c and Δq may be useful solute structural descriptors for retention prediction. As these indices quantify nonspecific and polar contributions to retention, their description in terms of nonempirical structural parameters would be desired. Once this is done, the Jandera's approach [74–77] would be useful in predicting of relative retention of complex polyfunctional solutes of real analytical importance.

An interesting approach to predicting RP HPLC retention, employing the partial molar volumes of structural fragments and energies of interactions of bond dipoles with water, has recently been developed by Galushko [80–82]. The equation derived for calculating of capacity factor k' of a solute X is

$$\ln k'_X = a\left(\sum_i V_i\right)^{2/3} + b\left(\sum_j \Delta G_{es.j,w}\right) + c, \tag{8.28}$$

where V_i are the increments of the partial molar volumes of structural fragments in water; $\Delta G_{es.j,w}$ are the increments of energy of electrostatic interaction of bond dipoles with water; a, b and c are constants characterizing individual RP HPLC systems (*i.e.*, stationary phase and mobile phase). Having several reference compounds of known volumes of fragments and bond dipole moments, the coefficients a, b and c can be calculated by regression analysis. That way the RP HPLC columns can be calibrated and different separation systems compared.

The starting point for Galushko method [80–82] was the solvophobic theory of RP HPLC [21]. The assumptions and simplifications suggested by Galushko [80] seem to be acceptable and convincing. The agreement between the actual and the calculated retention parameters for a set of aromatic derivatives [80] is perhaps not impressive but the model is consistent and (at least semiquantitatively) accounts for changes in main variables of the RP HPLC systems.

To calculate the values of $\sum V_i$ required for Eq. (8.28) Galushko [80] used experimental partial molar volume data collected by Belousov [83] for more than

300 compounds. Such a large collection of experimental molar volumes suffices for calculation of V_i for majority of fragments. If there is no appropriate experimental data the fragment volumes are simply derived from Van der Waals volumes.

The energy of interaction of solute with surrounding medium was calculated considering the solute molecule as consisting of dipoles, each of which separately interacts with the medium. Bond dipole moments are employed for calculation of the second term in Eq. (8.28). Simple arithmetics is applied because the bond dipole moments are known for almost all bonds.

Volume and energy increments for about 100 structural fragments and about 30 bond dipoles have been presented [82]. The comparison of experimental RP HPLC $\ln k'$ data for simple aromatic solutes with the calculated ones provided by Galushko *et al.* [82] shows differences up to nearly 5 ln units. At present days, with the readily available solute structural descriptors from molecular modeling, one is probably able to derive better QSRR. The Galushko approach deserves attention because it helps to understand major molecular and submolecular factors affecting the RP HPLC.

8.4. QSRR Employing Solvatochromic Parameters

The linear solvation energy relationship (LSER) methodology has become over the past decade one of the most often applied technique in studies of solute-solvent interactions [84–86]. Originally derived from UV spectroscopy, its main parameters found application throughout spectroscopy, in chemical kinetic and equilibrium processes and in solute transfer phenomena such as solubility, partitioning and chromatographic separation [87–89].

In first QSRR papers involving solvatochromic parameters as descriptors of RP HPLC retention the authors [90,91] proposed the following starting equation:

$$\log k' = \text{constant} + M(\delta_m^2 - \delta_s^2)V_2/100 + S(\pi_s^* - \pi_m^*)\pi_2^* + A(\beta_s - \beta_m)\alpha_2 + B(\alpha_s - \alpha_m)\beta_2, \tag{8.29}$$

where subscript 2 designates a solute property such as molar volume (V_2) polarizability-dipolarity (π_2^*), hydrogen bond acidity (α_2) and hydrogen bond basicity (β_2). Each solute property is multiplied by a term that represents the difference in complementary "solvent" properties of the mobile phase (subscript m) and the stationary phase (subscript s). Thus, α_m and α_s are the abilities of the phases (bulk or bonded) to donate a hydrogen bond. These properties complement the solute's ability to accept a hydrogen bond (β_2). Similarly, δ_m^2 and δ_s^2, the squares of the Hildebrand solubility parameter or cohesive energies of the two phases, complement the solute molar volume. In Eq. (8.29) M, S, A and B are constants.

In QSRR studies of retention data obtained for a given set of solutes at the same chromatographic conditions Eq. (8.29) reduces to a form

$$\log k' = c_0 + c_1 V_2/100 + c_2\pi_2^* + c_3\alpha_2 + c_4\beta_2, \tag{8.30}$$

where c_0–c_4 are regression coefficients.

Recently, Abraham [92] proposed the following modification of Eq. (8.30) (valid for all properties of analytes, including chromatographic retention):

$$\log \mathrm{SP} = c_0 + c_1 V_X + c_2 \pi_2^{\mathrm{H}} + c_3 \alpha_2^{\mathrm{H}} + c_4 \beta_2^{\mathrm{H}} + c_5 R_2, \tag{8.31}$$

where SP means some measurable property of a series of solutes in a fixed system; V_X is McGovan's characteristic volume [93] calculatable simply from molecular structure; π_2^{H} is the solute dipolarity/polarizability which can be determined through gas-chromatographic and other measurements; α_2^{H} is the effective or summation hydrogen-bond acidity; β_2^{H} is the effective or summation hydrogen-bond basicity and R_2 is an excess molar refraction [87] which can be obtained from refractive index measurements and is an additive quantity.

Carr *et al.* [91] analysed $\log k'$ data from RP HPLC determined for a set of benzene derivatives in five acetonitrile-water systems at four different temperatures. By regression analysis they derived coefficients c_0–c_4 of 20 equations of the form of Eq. (8.30). One can note that with the data set considered, not every variable of Eq. (8.30) appears statistically significant. The meaningful equations obtained contained the $V_2/100$, β_2 and either π_2^* or α_2 terms. The authors [91] opted for π_2^* as the third regression parameter. The predictive efficiency of the three-variable regression equations obtained was very high. It should be noted, however, that the authors [91] excluded several outlier solutes (for reasons not explained) when deriving their individual regression equations. The highest number of solutes considered in the regression was 21 out of total of 26. For nine of the reported equations, the number of solutes taken into consideration was only 15–17 out of total of 20.

The statistical significance of individual regression terms of Eq. (8.30) was critically analysed in case of retention data of benzodiazepines determined in various normal-phase and reversed-phase HPLC and TLC systems [94,95]. The authors [95] concluded regarding reversed-phase HPLC and TLC that when methanol was used as an organic modifier, the leading term in the correlation equations was $V_2/100$. This term could explain more than 99% of the total variance of retention parameters, the other terms being in most instances zero and not statistically significant at the 95% confidence level. Also, the term relating to solute hydrogen bonding basicity was significant at the 99% confidence level in those equations where acetonitrile and tetrahydrofuran were the organic modifiers. In case of normal-phase systems, in comparison with the reversed-phase systems, the influence on retention of the $V_2/100$ term was much less important and not always statistically significant at the 99% confidence level.

The findings by the Italian group are consistent with the common view regarding the differences in mechanism of the normal-phase and reversed-phase chromatographic separations. The formal shortages of the QSRR analysis presented by the group are: very limited size of the series of test solutes (only 8 compounds) and low variability of their solvatochromic parameters (*e.g.*, there were only two values 0 and 0.33 for α_2).

In their comparative studies of reversed-phase retention behaviour of polystyrene-divinylbenzene (PS-DVB) and alkyl-bonded silica stationary phases Pedigo and

Bowers [96] omitted the term $c_3\alpha_2$ in Eq. (8.30). The greatest difference was seen in c_2, the coefficient for dipolarity/polarizability term. The coefficient c_2 was always found positive in case of PS-DVB because of the aromaticity of the phase. This indicates that a solute with a large π_2^* is relatively more retained on PS-DVB than a solute with a low π_2^*. In contrast to PS-DVB, alkyl-bonded silica phases showed c_2 values close to zero (with tetrahydrofuran and acetonitrile as eluent organic modifiers) or negative (with methanol as eluent modifier). This fact indicates that the polarizability and dipolarity of a solute are less important in retention on alkyl-bonded silica phases.

Chinese workers employed solvatochromic parameters in QSRR equations describing the slope of the relationship of $\log k'$ *versus* fraction of organic solvent in a binary aqueous mobile phase [97–99]. From the analysis of the significance of the coefficients of the individual regression variables, V_2, π_2^*, β_2 and α_2, it is evident that the term representing hydrogen bonding acidity, α_2, is statistically insignificant. This is in spite of using a sufficiently large group of solutes ($n = 49$) for the QSRR study. Thus, these authors obtained basically the same form of relationship as the ones discussed above.

Park *et al.* [100] related $\log k'$, determined for eight small aromatic solutes on octadecyl-silica with methanol-water as the eluent, to the V_2 and β_2 parameters. The small size of the solute series limited the possibility of demonstrating significance for RP HPLC of other variables in Eq. (8.30). However, the recent paper by Park *et al.* [101] brings also rather unexpected results. These authors determined $\log k'$ for 24 benzene and pyridine derivatives on several silica-based stationary phases (C_8, C_{18}, diphenylmethyl, cyanopropyl) with methanol-buffer and acetonitrile-buffer eluents. With aqueous methanol eluent, all the QSRR equations derived comprised no significant dipolarity/polarizability term. In case of acetonitrile-water eluent, only the QSRR obtained for $\log k'$ from cyanopropylsilica columns had no $c_2\pi_2^*$ term. Does it really mean that polar interactions are unimportant for HPLC retention of small aromatic solutes? Alternative explanation is that the input to $\log k'$ by these interactions is similar for all the solutes studied and is accounted for by the constant c_0 in Eq. (8.30). As suggested by Park *et al.* [101] the magnitude of coefficients c_0–c_4 of Eq. (8.30) may be used to quantitatively compare stationary phase materials for HPLC similarly to the Rohrschneider and McReynolds constants used to characterize gas-liquid chromatographic phases.

Based on their extensive study of RP HPLC capacity factors extrapolated to 100% water as an eluent, Hsieh and Dorsey [102] conclude that the most important solute parameters which influence retention are solute size and hydrogen bonding basicity, but not hydrogen bonding acidity. The solute dipolarity appears a minor but still significant factor. Exactly the same conclusion was drawn by Altomare *et al.* [103] from their QSRR employing solvatochromic parameters of 71 solutes chromatographed on a C_8 silica-based HPLC stationary phase.

Comparative solvatochromic analysis of RP HPLC data and logarithms of *n*-octanol-water partition coefficients ($\log P$) was done by Vallat *et al.* [104] for about 40 diversified solutes. It was concluded that an octadecyl polyvinyl-alcohol copolymer (ODP) stationary phase provided retention data closely resembling partition in the *n*-octanol-water systems.

To the solvatochromic analysis of RP HPLC data, from the point of view of relationship to $\log P$, is devoted a recent series of papers by Abraham and co-workers [105–107].

It can be summarised that solvatochromic parameters are very helpful in understanding the factors determining retention in individual chromatographic systems. By means of these parameters one can rationalize the differences in separation of solutes observed in normal-phase and reserved-phase chromatography. Also, the separation specificity of individual stationary phases seems to be properly accounted for by the coefficients at solvatochromic regressors in QSRR equations. It is still difficult, however, to judge whether solvatochromic parameters are more suitable and convenient for the description of retention of chemical compounds than the parameters obtained theoretically by molecular modeling and computational chemistry. Solvatochromic parameters are empirical. Respective data are available for an increasing number of compounds [87,89]. On the other hand, theoretical descriptors can be obtained readily for any structure owing to the common access to calculation chemistry software.

8.5. QSRR Based on Physical Properties of Solute Substances

The solubility parameter theories of chromatographic separations [108,109] provide basis for predicting retention. Based on several publications [110–117] it can be concluded that the extended solubility parameter (ESP) model is very capable of predicting, at least, the trends in retention behaviour of solutes. There are extensive collections of solubility parameters [118,119]. Tijssen *et al.* [117] offer such data for about 200 compounds of chromatographic interest. These data are of utmost suitability for QSRR aimed at optimization of separations and elucidation of retention mechanism operating in a given chromatographic system. However, when it comes to prediction of retention of new solutes the use of theoretical and semiempirical approaches to estimate the solubility parameters is difficult because the proposed equations are very complicated. In view of that interesting appears a simplified approach to evaluate the total solubility parameter, δ_T^2, by a group of Chinese authors [120–122].

The authors [122] derived regression equation predicting δ_T^2 based on solute molar volume, V, and boiling point, T_b:

$$\delta_T^2 = -2.085 \times 10^{-2}\frac{T_b^2}{V} + 58.93\frac{T_b}{V} - 6892.14\frac{1}{V} - 26.76. \qquad (8.32)$$

Eq. (8.32) was derived for 60 common compounds. According to authors [120–122] it provided “satisfactory GC and LC retention data”. Unfortunately, the authors do not supply any information on statistical significance of individual terms in Eq. (8.32) nor on intercorrelations among them. The question arises if Eq. (8.32) is able correctly predict retention of a solute which has not been used to derive regression.

Solute boiling point, T_b, has occasionally been tested as retention predictor in GC. Leaving aside homologous series of solutes, its suitability for QSRR appears limited.

One of the first QSRR involving T_b was by Dimov and Boneva [123]. These authors proposed the following equation to describe relative GC retention, R_i, of seven 2-nitroalkanol-1 compounds on *Carbowax* 20*M* phase:

$$R_i = -9.70 + 0.044T_b + 0.493\frac{M}{(W + h_{NO_2})}, \tag{8.33}$$

where M is solute molecular mass; W is the Wiener number and h_{NO_2} is a measure of NO_2 hindrance to which a value of 20 units has been assigned for tertiary NO_2 groups and 10 units otherwise. Even if there is no chance correlation, obtained with four independent variables and only seven dependent variable data points, it is difficult to assign any physical meaning to the last term in Eq. (8.33).

Interesting appears analysis of the QSRR equations reported by Bermejo and co-workers [124–127]. In regression equations the solute boiling point and some molecular bulkiness parameters were the independent variables predicting Kováts indices determined on phases of varying polarity. The bulkiness parameters considered were molar volume, molar refractivity and first-order molecular connectivity as well as their reciprocals. The solutes analysed were isomeric hydrocarbons and chlorinated 1,4-dimethylbenzenes. The equations obtained were of the general form:

$$I = c_0 + c_1 T_b + c_2 \text{ Bulk}, \tag{8.34}$$

where I was the retention index, T_b was the solute boiling point and *Bulk* was molar volume, molar refractivity or molecular connectivity (or their reciprocals); c_0–c_2 were regression coefficients. From a formal statistical point of view several of the reported QSRR equations are disputable because of the evident intercorrelations between T_b and *Bulk* variables. There has been an attempt, however, to interpret the empirical relationships expressed by Eq. (8.34) in fundamental physicochemical terms [128].

Experimentally determined total dipole moment has occasionally been employed in multiparameter QSRR equations although the respective parameter from molecular modeling is normally preferred (see Chapter 9). In one report [129] a three-parameter regression equation was derived containing the square of the dipole moment, Van der Waals volume and the hydrogen bond energy as independent variables. The dependent variable was the coefficient of an equation derived by the authors and relating ln k' from HPLC to changes in mobile phase composition. There is no information in the original paper on the significance of individual regressors in QSRR equations.

According to Ong and Hites [130] predictive QSRR can be obtained regarding retention indices of polyaromatic hydrocarbons, chlorinated biphenyls, dibenzo-*p*-dioxins and dibenzofurans by linear combination of molecular polarizability, the ionization potential and the square of dipole moment of solutes.

When dealing with the empirically determined chromatographic retention descriptors one should not forget about the fundamental Rohrschneider approach [131] to retention prediction. The essence of the concept is that under the conditions of classical GC, where the column temperature, carrier gas inlet and outlet pressures as well as its flow rate are constant during the analysis, the individual interactions of any substance with a given phase can be described by means of interactions of a few selected reference substances (*e.g.*, benzene, ethanol, methyl ethyl ketone, nitromethane and pyridine) and the specific constants of the examined substance (Rohrschneider's constants). The difference of the retention index of a given solute on a given stationary phase and on the reference *Squalane* phase, ΔI_S, is obtained as a sum of the products of the stationary phase specific and solute specific factors. The concept was further developed by McReynolds [132]. The approach to retention prediction resulting from works of Rohrschneider, McReynolds, Ladon and Sandler [133] and many others may be of practical value if the solute and the stationary phase constants are known. Unfortunately, this is not often the case when QSRR for a given series of solutes is required.

8.6. Quantitative Structure-Electrophoretic Mobility Relationships

Semiquantitative relationships between electrophoretic mobility and molecular mass of solutes have been discussed in section 5.5. First regular QSRR regarding electrophoretic data has been reported very recently [134]. Considering migration time data of Chen and Cassidy [135] for a series of 13 lanthanides Kaliszan *et al.* [134] found good correlations between reciprocal of migration time, $1/t$ and both atomic mass, A_m, and ionic radius, I_r (the two structural descriptors being highly intercorrelated). The correlations observed may appear trivial in case of such closely congeneric solutes like lanthanides. However, a single structure descriptor, like A_m or I_r, did not suffice to account for the differences in migration time of a more diverse set of metals like the series listed in Table 8.1 [136]. Inadequacy of a single descriptor to account for migration differences of the whole set of cations is illustrated in Fig. 8.2 for I_r.

The following multiple regression equations were derived:

$$1/t = 0.09(\pm 0.04) + 0.342(\pm 0.040)I_r - 0.0016(\pm 0.0001)A_m,$$
$$p \leq 0.01 \qquad p \leq 10^{-4} \qquad p \leq 10^{-4}$$
$$n = 26 \quad R = 0.940 \quad s = 0.03 \quad p \leq 10^{-4}; \tag{8.35}$$

$$1/t = -0.14(\pm 0.08) + 0.442(\pm 0.047)I_r - 0.0015(\pm 0.0001)A_m + 0.00023(\pm 0.00007)E_i,$$
$$p \leq 0.1 \qquad p \leq 10^{-4} \qquad p \leq 10^{-4} \qquad p \leq 6 \times 10^{-3}$$
$$n = 26 \quad R = 0.961 \quad s = 0.03 \quad p \leq 10^{-4}, \tag{8.36}$$

where E_i is energy of ionization; p denotes the significance levels of the individual regression terms and of the whole equation; n is the number of data points used in regression analysis; R is the correlation coefficient; s is the standard error of estimate. Numbers in parenthesis are standard deviations of regression coefficients. Intercorrelations between the pairs of descriptors used in Eqs. (8.35) and (8.36) are: I_r *vs* A_m $r = 0.32$; I_r *vs* E_i $r = 0.73$ and A_m *vs* E_i $r = 0.37$. Thus, the intercorrelation between I_r and A_m is negligible for the series of solutes studied. Also, the correlation between I_r and E_i is low enough to keep the two variables in the same regression

Table 8.1 Electrophoretic migration time after Chen and Cassidy [136], atomic mass, ionic radius and energy of ionization of a series of metal cations.

Cation	*Migration time* $t_m[min]$	*Atomic mass* A_m	*Ionic radius* $I_r[\AA]$	*Energy of ionization* $E_i[kJ/mol]$
K	2.1	39.1	1.33	419
Ba	2.6	137	1.34	503
Ca	2.8	40.1	0.99	590
Na	3.0	23.0	0.97	496
Mg	3.1	24.3	0.66	738
Mn(II)	3.2	54.9	0.80	718
Fe(II)	3.4	55.8	0.74	761
Co(II)	3.4	58.9	0.72	757
Ni(II)	3.6	58.7	0.69	735
Zn	3.7	95.4	0.74	906
Li	3.8	6.94	0.68	520
La	3.9	139	1.06	538
Ce	4.3	140	1.03	528
Pr	4.5	141	1.01	522
Nd	4.6	144	1.00	530
Sm	5.2	150	0.96	542
Eu	5.5	152	0.95	547
Gd	5.7	157	0.94	595
Cu(II)	6.1	63.5	0.72	745
Tb	6.4	159	0.92	569
Dy	6.8	162	0.91	567
Ho	7.3	165	0.89	574
Er	7.8	167	0.88	581
Tm	8.5	169	0.87	588
Yb	9.3	173	0.93	603
Lu	9.9	175	0.85	513

equation. Correlations of individual descriptors *versus* pairs of other descriptors are: I_r *vs* $A_m + E_i$ $R = 0.73$; A_m *vs* $I_r + E_i$ $R = 0.38$ and E_i *vs* $I_r + A_m$ $R = 0.74$.

The above example demonstrates the high predictive potency of QSRR in case of inorganic cations. Having an electropherogram of a mixture of cations with several metals known one can derive a QSRR equation and then identify the remaining peaks by means of basic parameters taken from a periodic table.

A brief discussion deserves the physical meaning of the statistically derived relationships. As evident from basic electrophoretic equation (Eq. (5.7)) there are two solute dependent variables which affect electrophoretic mobility (or migration time) determined in the same separation system. These are: the effective charge, z, and the radius of spherical particles, a, playing a role of retardation factor. The structural parameters present in Eq. (8.36) were the "best" selected from a variety of descriptors tested. For example, among others, insignificant for description of migration appeared the nominal ionic charge of the cations. The successful prediction of relative electrophoretic behaviour of metal cations by QSRR indicates that atomic mass approximates the retardation factors (negative input to $1/t$) whereas the ionic radius is a more or less approximate measure of the effective charge. Energy of ionization can play a role of a secondary (however significant) correction factor to the effective charge. A higher electrophoretic mobility have cations with high ionization energies (cations with low polarizability: the so called "hard" cations). Their interactions with

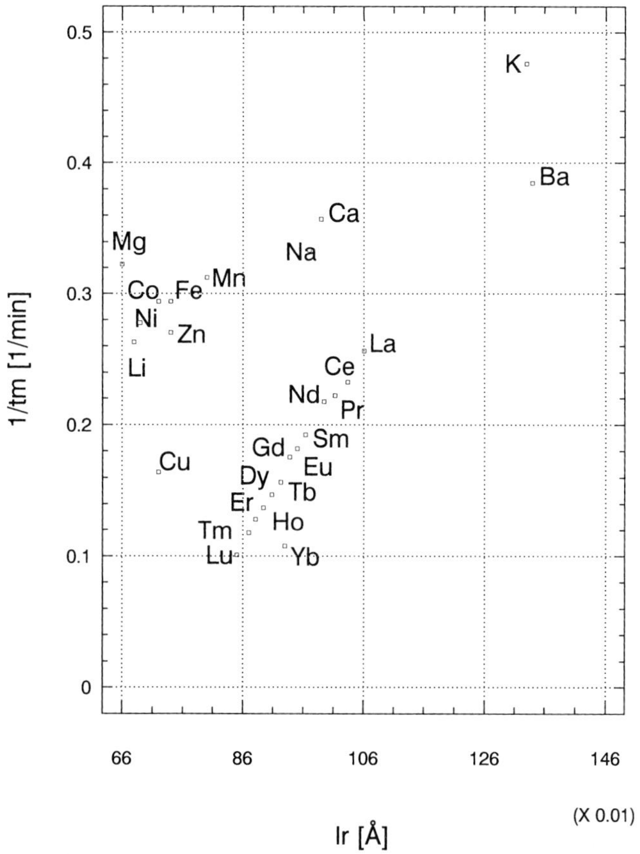

Figure 8.2. Distribution of metal cations on the plane determined by reciprocal of electrophoretic migration time according to Chen and Cassidy [136] and ionic radius. (Reprinted with permission from Kaliszan, R., Turowski, M., Buciński, A. and Hartwick, R.A., *Quant. Struct.-Act. Relat.* 14, 356, 1995. Copyright VCH Verlagsgesellschaft mbH.)

the environment (which decrease electrophoretic mobility) are weaker than in case of more polarizable cations.

The attempts [134,137] to derive QSRR in case of electrophoretic data of organic solutes cannot be called successful. Certainly, contrary to spherical inorganic cations, irregular shape of organic ions affects mobility in a manner difficult to quantify. It was demonstrated [138] that molecular orientation of rod-shaped organic ions plays a significant role in determining electrophoretic mobility. The description is only semi-quantitative. The challenge, however, is to find an organic equivalent to the ionic radius of metal cations which would be a reliable descriptor of effective charge.

References

1. Evans, M.B. and Haken, J.K., *J. Chromatogr.* **472**, 93 (1989).
2. Peng, C.T., *J. Chromatogr. A* **678**, 189 (1994).

3. Ono, A., *J. Chromatogr.* **110**, 233 (1975).
4. Haken, J.K., Nguyen, A. and Wainwright, M.S., *J. Chromatogr.* **178**, 471 (1979).
5. Jinno, K. and Kawasaki, K., *Chromatographia* **18**, 90 (1984).
6. Yamagami, C., Takami, H., Yamamoto, K., Miyoshi, K. and Takao, N., *Chem. Pharm. Bull.* **32**, 4994 (1984).
7. Zou, H.F., Wang, Q.S., Gao, R.Y., Yang, H.Z., Yang, B.W., Zhang, Y.K. and Lu, P.C., *Chromatographia* **31**, 143 (1991).
8. Kim, H.S., Kim, T.K. and Lee, D.W., *J. Liq. Chromatogr.* **17**, 2615 (1994).
9. Martin, A.J.P., *Biochem. Soc. Symp.* **3**, 4 (1950).
10. Green, J., Marcinkiewicz, S. and McHale, D., *J. Chromatogr.* **10**, 158 (1963).
11. Sissons, D. and Welti, D., *J. Chromatogr.* **60**, 15 (1971).
12. Tejedor, J.N., *J. Chromatogr.* **177**, 279 (1979).
13. Budahegyi, M.V., Lombosi, E.R., Lombosi, T.S., Mészaros, S.Y., Nyiredy, S., Tarján, G., Timár, I. and Takács, J.M., *J. Chromatogr.* **271**, 213 (1983).
14. Mihara, S. and Enomoto, N., *J. Chromatogr.* **324**, 428 (1985).
15. Osiek, J., Oszczapowicz, J. and Drzewiński, W., *J. Chromatogr.* **351**, 177 (1986).
16. Chen, B.-K. and Horváth, C., *J. Chromatogr.* **171**, 15 (1979).
17. Kuchař, M., Rejholec, V., Jelinková, M. and Nemeček, O., *J. Chromatogr.* **150**, 419 (1978).
18. DiBussolo, J.M. and Nes, W.R., *J. Chromatogr. Sci.* **20**, 193 (1982).
19. Riley, C.M., Tomlinson, E. and Jefferies, T.M., *J. Chromatogr.* **185**, 197 (1979).
20. Tomlinson, E., Poppe, H. and Kraak, J.C., *Int. J. Pharm.* **7**, 225 (1981).
21. Horváth, C., Melander, W. and Molnar, I., *J. Chromatogr.* **125**, 129 (1976).
22. Morishita, F., Kakihana, H. and Kojima, T., *Anal. Lett.* **17A**, 2385 (1984).
23. Smith, R.M. and Burr, C.M., *J. Chromatogr.* **475**, 57 (1989).
24. Colin, H., Krstulovic, A.M., Gonnord, M.-F., Guiochon, G., Yun, Z. and Jandera, P., *Chromatographia* **19**, 9 (1983).
25. Smith, R.M. and Wang, R., *J. Chromatogr.* **558**, 7 (1991).
26. Smith, R.M. and Burr, C.M., *J. Chromatogr.* **481**, 71 (1989).
27. Gilpin, R.K., Jaroniec, M. and Lin, S., *Chromatographia* **30**, 393 (1990).
28. Alvarez-Zepeda, A., Barman, B.N. and Martire, D.E., *Anal. Chem.* **64**, 1978 (1992).
29. Smith, R.M., *J. Chromatogr. A* **656**, 381 (1993).
30. Molnár, and Horváth, C., *J. Chromatogr.* **145**, 371 (1978).
31. Gill, R., Alexander, S.P. and Moffat, A.C., *J. Chromatogr.* **218**, 639 (1981).
32. Huang, J.-X., Bouvier, E.S.P., Stuart, J.D., Melander, W.R. and Horváth, C., *J. Chromatogr.* **330**, 181 (1985).
33. Assenza, S.P. and Brown, P.R., *J. Chromatogr.* **282**, 477 (1983).
34. Walczak, B., Dreux, M., Chretien, J.R., Szymoniak, K., Lafosse, M., Morin-Allory, L. and Doucet, J.P., *J. Chromatogr.* **353**, 109 (1986).
35. Głowniak, K. and Bieganowska, M.L., *J. Chromatogr.* **370**, 281 (1986).
36. Gago, F., Alvarez-Builla, J. and Elguero, J., *J. Chromatogr.* **449**, 95 (1988).
37. Dondi, F., Blo, G., Kahie, Y.D., Lodi, G., Pietrogrande, C. and Reschiglian, P., *Chromatographia* **25**, 423 (1988).
38. Dondi, F., Kahie, Y.D., Lodi, G., Blo, G., Pietrogrande, C. and Reschiglian, P., *J. Chromatogr.* **461**, 281 (1989).
39. Yamagami, C. and Takao, N., *Chem. Pharm. Bull.* **40**, 925 (1992).
40. Pietrogrande, M.C., Reschiglian, P., Dondi, F., Kahie, Y.D. and Bertolasi, V., *J. Chromatogr.* **592**, 65 (1992).
41. Valko, K., Szabo, G., Röhricht, J., Jemnitz, K. and Darvas, F., *J. Chromatogr.* **485**, 349 (1989).
42. Tsantili-Kakoulidou, A., El Tayar, N., van de Waterbeemd, H. and Testa, B., *J. Chromatogr.* **389**, 33 (1987).
43. Brändstrom, A., *Acta Pharm. Scand.* **19**, 175 (1982).
44. Fujita, T., *Progr. Phys. Org. Chem.* **14**, 75 (1983).
45. Smith, R.M. and Burr, C.M., *J. Chromatogr.* **550**, 335 (1991).
46. Spanjer, M.C. and De Ligny, C.L., *J. Chromatogr.* **253**, 23 (1982).
47. Spanjer, M.C. and De Ligny, C.L., *Chromatographia* **20**, 120 (1985).
48. Mihara, S. and Masuda, H., *J. Chromatogr.* **402**, 309 (1987).
49. Tanaka, N., Goodell, H. and Karger, B.L., *J. Chromatogr.* **158**, 233 (1978).
50. Wells, M.J.M. and Clark, C.R., *Anal. Chem.* **64**, 1660 (1992).
51. Lurie, I.S. and Allen, A.C., *J. Chromatogr.* **292**, 283 (1984).
52. Smith, R.M., Murilla, G.A., Hurdley, T.G., Gill, R. and Moffat, A.C., *J. Chromatogr.* **384**, 259 (1987).
53. Smith, R.M. and Garside, D.R., *J. Chromatogr.* **407**, 19 (1987).
54. Sun, J.J. and Fritz, J.S., *J. Chromatogr.* **522**, 95 (1990).

55. Hafkenscheid, T. and Tomlinson, E., *J. Chromatogr.* **218**, 409 (1981).
56. Altomare, C., Carotti, A., Cellamare, S. and Ferappi, M., *Int. J. Pharm.* **56**, 273 (1989).
57. Fredenslund, A., Jones, R.L. and Prausnitz, J.M., *AIChE J.* **21**, 1086 (1975).
58. Fredenslund, A., Gmehling, J. and Rasmussen, P., "*Vapor-Liquid Equilibria using UNIFAC*", Elsevier, Amsterdam, 1977.
59. Abrams, D.S. and Prausnitz, J.M., *AIChE J.* **21**, 116 (1975).
60. Derr, E.L. and Deal, C.H., *Inst. Chem. Eng. Symp. Ser., No. 32*, **3**, 40 (1969).
61. Guggenheim, E.A., "*Mixtures*", Clarendon Press, Oxford, 1952.
62. Gmehling, J., Rasmussen, P. and Fredenslund, A., *Ind. Eng. Chem., Process Des. Dev.* **21**, 118 (1982).
63. Roth, M. and Novak, J., *J. Chromatogr.* **258**, 23 (1983).
64. Petrovic, S.M., Lomic, S. and Sefer, I., *J. Chromatogr.* **348**, 49 (1985).
65. Gmehling, J. and Weidlich, U., *Fluid Phase Equilibria* **27**, 171 (1986).
66. Petrovic, S.M. and Lomic, S., *J. Liq. Chromatogr.* **12**, 59 (1989).
67. Price, G.J. and Dent, M.R., *J. Chromatogr.* **483**, 1 (1989).
68. Dasko, L., *J. Chromatogr.* **543**, 267 (1991).
69. Park, J.H., Lee, J.E., Jang, M.D., Li, J. and Carr, P.W., *J. Chromatogr.* **586**, 1 (1991).
70. Price, G.J. and Dent, M.R., *J. Chromatogr.* **585**, 83 (1992).
71. Price, G.J., *J. Chromatogr.* **586**, 297 (1992).
72. Park, J.H., Jang, M.D., Chae, J.J., Kim, H.C. and Suh, J.K., *J. Chromatogr. A* **656**, 69 (1993).
73. Jandera, P., Colin, H. and Guiochon, G., *Anal. Chem.* **54**, 435 (1982).
74. Jandera, P., *J. Chromatogr.* **314**, 13 (1984).
75. Jandera, P., *J. Chromatogr.* **352**, 91 (1986).
76. Jandera, P. and Kubál, J., *J. Chromatogr.* **500**, 281 (1990).
77. Jandera, P., *J. Chromatogr. A* **656**, 437 (1993).
78. Hansch, C. and Leo, A., "*Substituent Constants for Correlation Analysis in Chemistry and Biology*", Wiley, New York, 1979.
79. Snyder, L.R. and Kirkland, J.J., "*Introduction to Modern Liquid Chromatography*", Wiley, New York, 1979.
80. Galushko, S.V., *J. Chromatogr.* **552**, 91 (1991).
81. Galushko, S.V., *Chromatographia* **36**, 39 (1993).
82. Galushko, S.V., Kamenchuk, A.A. and Pit, G.L., *J. Chromatogr. A* **660**, 47 (1994).
83. Belousov, V.P., "*Termodinamika Vodnykh Rastvorov Neelektrolitov*", Khimya, Leningrad, 1983.
84. Kamlet, M.J., Abboud, J.-L.M. and Taft, R.W., *Prog. Phys. Org. Chem.* **13**, 485 (1981).
85. Abraham, M.H., Doherty, R.M., Kamlet, M.J. and Taft, R.W., *Chem. Br.* **22**, 551 (1986).
86. Kamlet, M.J., Doherty, R.M., Abbound, J.-L.M., Abraham, M.H. and Taft, R.W., *CHEMTECH* **1986**, 566 (1986).
87. Abraham, M.H., Whiting, G.S., Doherty, R.M. and Shuely, W.J., *J. Chem. Soc. Perkin Trans.* 2, **1990**, 1451 (1990).
88. Leahy, D.E., Morris, J.J., Taylor, P.J. and Wait, A.R., *J. Chem. Soc. Perkin Trans.* 2, **1992**, 705 (1992).
89. Abraham, M.H., Chadha, H.S., Whiting, G.S. and Mitchell, R.C., *J. Pharm. Sci.* **83**, 1085 (1994).
90. Sadek, P.C., Carr, P.W., Doherty, R.M., Kamlet, M.J., Taft, R.W. and Abraham, M.H., *Anal. Chem.* **57**, 2971 (1985).
91. Carr, P.W., Doherty, R.M., Kamlet, M.J., Taft, R.W., Melander, W. and Horváth, C., *Anal. Chem.* **58**, 2674 (1986).
92. Abraham, M.H., *Chem Soc. Rev.* **22**, 73 (1993).
93. Abraham, M.H. and McGovan, J.C., *Chromatographia* **23**, 243 (1987).
94. Pietrogrande, M.C., Dondi, F., Blo, G., Borea, P.A. and Bighi, C., *J. Liq. Chromatogr.* **11**, 1313 (1988).
95. Pietrogrande, M.C., Bighi, C., Borea, P.A. and Dondi, F., *J. Chromatogr.* **522**, 37 (1990).
96. Pedigo, S. and Bowers, L.D., *J. Chromatogr.* **499**, 279 (1990).
97. Zou, H., Zhang, Y. and Lu, P., *J. Chromatogr.* **522**, 49 (1990).
98. Zou, H., Wang, Q.S., Gao, R.Y, Yang, H.Z., Yang, B.W., Zhang, Y.K. and Lu, P.C., *Chromatographia* **31**, 143 (1991).
99. Chen, N., Zhang, Y. and Lu, P., *J. Chromatogr.* **606**, 1 (1992).
100. Park, J.H., Jang, M.D. anf Shin, M.J., *J. Chromatogr.* **595**, 45 (1992).
101. Park, J.H., Chae, J.J., Nah, T.H. and Jang, M.D., *J. Chromatogr. A* **664**, 149 (1994).
102. Hsieh, M.-M. and Dorsey, J.G., *J. Chromatogr.* **631**, 63 (1993).
103. Altomare, C., Cellamare, S., Carotti, A. and Ferappi, M., *Quant. Struct.-Act. Relat.* **12**, 261 (1993).
104. Vallat, P., Fan, W., El Tayar, N., Carrupt, P.-A. and Testa, B., *J. Liq. Chromatogr.* **15**, 2133 (1992).
105. Abraham, M.H. and Roses, M., *J. Phys. Org. Chem.* **7**, 672 (1994).
106. Abraham, M.H., Chadha, H.S. and Leo, A.J., *J. Chromatogr. A* **664**, 203 (1994).
107. Abraham, M.H., Chadha, H.S., Dixon, J.P. and Leo, A.J., *J. Phys. Org. Chem.* **7**, 712 (1994).

108. Tijssen, R., Billiet, H.A.H. and Schoenmakers, P.J., *J. Chromatogr.* **122**, 185 (1976).
109. Karger, B.L., Snyder, L.R. and Eon, C., *J. Chromatogr.* **125**, 71 (1976).
110. Schoenmakers, P.J., Billiet, H.A.H. and De Galan, L., *Chromatographia* **15**, 205 (1982).
111. Szanto, J. and Veres, T., *Chromatographia* **20**, 205 (1985).
112. Schoenmakers, P.J., "*Optimization of Chromatographic Selectivity, a Guide to Method Development*", Elsevier, Amsterdam, 1986.
113. Cooper, H.A. and Hurtubise, R.J., *J. Chromatogr.* **360**, 327 (1986).
114. Park, J.H. and Carr, P.W., *Anal. Chem.* **59**, 2596 (1987).
115. Smith, P.L. and Cooper, W.T., *Chromatographia* **25**, 55 (1988).
116. Sreenivasan, K., *J. Liq. Chromatogr.* **13**, 383 (1990).
117. Tijssen, R., Schoenmakers, P.J., Böhmer, M.R., Koopal, L.K. and Billiet, H.A.H., *J. Chromatogr. A* **656**, 135 (1993).
118. Hoy, K.L., "*Tables of Solubility Parameter*", 3rd ed., Union Carbide Corporation, South Charleston, 1975.
119. Barton, A.F.M., "*CRC Handbook of Solubility Parameters and Other Cohesion Parameters*", CRC Press, Boca Raton, Florida, 1983.
120. Zhang, H. and Hu, Z., *Chromatographia* **33**, 575 (1992).
121. Hu, Z.D. and Zhang, H.W., *J. Chromatogr. A* **653**, 275 (1993).
122. Sun, Z.L., Liu, M.C. and Hu, Z.D., *Chromatographia* **38**, 599 (1994).
123. Dimov, N. and Boneva, S., *J. Chromatogr.* **206**, 549 (1981).
124. Bermejo, J. and Guillen, M.D., *J. Chromatogr.* **318**, 187 (1985).
125. Bermejo, J., Blanco, C.G. and Guillen, M.D., *J. Chromatogr.* **331**, 237 (1985).
126. Bermejo, J., Blanco, C.G. and Guillen, M.D., *J. Chromatogr.* **351**, 425 (1986).
127. Diez, M.A., Guillen, M.D., Blanco, C.G. and Bermejo, J., *J. Chromatogr.* **508**, 363 (1990).
128. Kaliszan, R., "*Quantitative Structure-Chromatographic Retention Relationships*", Wiley, New York, 1987.
129. Lu, P., Zou, H. and Zhang, Y., *J. Chromatogr.* **509**, 171 (1990).
130. Ong, V.S. and Hites, R.A., *Anal. Chem.* **63**, 2829 (1991).
131. Rohrschneider, L., *J. Chromatogr.* **22**, 6 (1966).
132. McReynolds, W.O., *J. Chromatogr. Sci.* **8**, 685 (1970).
133. Ladon, A. and Sandler, S., *Anal. Chem.* **45**, 921 (1973).
134. Kaliszan, R. Turowski, M., Buciński, A. and Hartwick, R.A., *Quant. Struct.-Act. Relat.* **14**, 356 (1995).
135. Chen, M. and Cassidy, R.M., *J. Chromatogr.* **602**, 227 (1992).
136. Chen, M. and Cassidy, R.M., *J. Chromatogr.* **640**, 425 (1993).
137. Lukkari, P., Vuorela, H. and Riekkola, M.-L., *J. Chromatogr.* **655**, 317 (1993).
138. Grossman, P.D. and Soane, D.S., *Anal. Chem.* **62**, 1592 (1990).

9 Multiparameter QSRR with Solute Descriptors from Calculation Chemistry

9.1. Quantum Chemical and Molecular Mechanics Indices as Solute Structure Parameters

Before the formal QSRR methodology had been established, some chromatographers turned their attention to quantum chemistry. Initially, the simple extended Hückel method (EHT) or Del Re's quantum chemical empirical method were employed for calculations of quantum chemical indices of solutes. First tested in correlation studies to retention were such quantum chemical data like the π-electron density [1,2] and the sum of "effective" charges on principal individual atoms in the solute molecule [3,4]. Approximate character of the indices employed and lack of statistical rigour do not deprive these first reports of historical importance.

One of the first applications in QSRR of quantum chemical structural descriptors calculated by the well known CNDO/2 method was specifically concerned with solute dipole moments [5]. The attempt by the authors [5] to describe GC retention indices of phenol derivatives in terms of experimental dipole moments has met with only limited success. Also, with the CNDO/2-calculated dipole moments the QSRR obtained were not impressive.

In some instance, choosing of proper solute conformation appeared to be a problem. This was demonstrated with 4-hydroxyphenol (Fig. 9.1). Among the conformations of the compound, two energetically favoured conformers exist. In such a situation the calculated dipole moment is obtained as the arithmetic mean of the two conformers and equals 1.61 D. However, such an average calculated dipole moment did not correlate with GC retention data. On the other hand, if one takes the dipole moment calculated for one conformer ($\mu = 3.21\,D$), there is a very good correlation with chromatographic data. Unfortunately, there is no *a priori* rationalization of such a choice.

Extensive QSRR studies on GC retention indices for a group of solutes consisting of aliphatic esters, ethers, alcohols, ketones and aldehydes were published in 1983 by Buydens *et al.* [6,7]. The authors used in multiple regression analysis the following

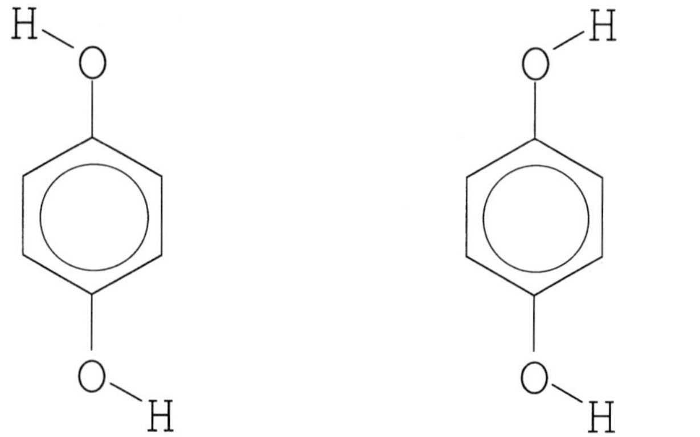

E_{total} = -34395.7565 kcal/mol E_{total} = -34395.7024 kcal/mol

μ = 0.000 D μ = 2.266 D

Figure 9.1. Dipole moments, μ, and conformational energies, E, for energetically favoured conformations of 4-hydroxyphenol.

CNDO/2-calculated electronic parameters: the magnitude of the dipole moment, DM; the sum of the absolute values of the charges in a given molecule, QT; and the sum of the absolute values of the charges of the atoms constituting the functional group and the atoms in the α position to this functional group, QA. The square terms of the individual structural parameters were also included in the regression analysis.

To get significant QSRR, the authors [6,7] had to consider separately the subsets of mono- and bifunctional derivatives. Multiparameter regression equations involving the quantum chemical indices, along with several topological indices and substituent constants, have shown the importance of electronic parameters for solute retention on polar stationary phases. Whereas for the nonpolar phases the topological parameters of monofunctional molecules could explain a great part of the total retention variance, for the polar stationary phases a combination of topological and electronic parameters was necessary. When the quantum chemical indices were included in the regression, they were preferred to the Hammett sigma.

There is a significant intercorrelation among the structural parameters used simultaneously in individual regression equations reported. This makes it difficult to interpret the physical meaning of a particular parameter applied. Nonetheless, one can agree with the authors [6,7] that the parameters describing local polarity (QA) are much more important than the total dipole moment (DM). This is especially evident when the retention indices on polar stationary phases are considered.

Another early QSRR study employing CNDO/2 indices was by Garcia-Raso *et al.* [8]. The GC retention data analysed were taken from literature [9] and concerned a series of *n*-alkenes chromatographed on *Squalane*. The regressors in the three-parameter equation were the total energy, E_T, the so-called solute-stationary phase interaction factor, ΔE, and the energy of the highest occupied molecular orbital, E_{HOMO}. The interaction factor, ΔE, was calculated from E_{LUMO} of solutes and E_{HOMO} of the stationary phase. The subgroups of 62 *cis* and *trans* alkenes had to be considered separately. The question arises whether the introduction of individual descriptors into regression equations is statistically justified or if a chance correlation takes place. In any case, the quantum chemically calculated total energy (or binding energy as well) is the parameter of deciding importance for retention of this closely congeneric group of solutes of relatively low and nearly equal polarity. This is the case for retention on nonpolar stationary phases.

It appeared, in studies on GC indices of a structurally diverse set of primary, secondary and tertiary (heterocyclic) amines determined on the methysilicone (OV-101) phase of low polarity that of several CNDO/2-calculated indices the total energy, E_T, had prevailing effect on retention [10]. Assuming that E_T reflects differences in nonpolar (dispersive) interactions of solutes with components of chromatographic systems, attempts were undertaken to characterise also a solute polarity increment to retention. However, when the CNDO/2-calculated dipole moment was used along with E_T in a two-parameter regression equation, its significance level was meaningless. To find a more reliable measure of chromatographic polarity the electron excess charges on individual atoms in a molecule were considered. In effect, a submolecular polarity parameter Δ has been proposed.

To determine Δ, electron densities on each atom in the molecule are determined and two atoms are found with highest electron excess and deficiency. The difference in electron densities for the two atoms gives Δ (Fig. 9.2).

In the two-parameter regression equation derived by the authors [10] and relating the Kováts index of the amines to E_T and Δ, the term Δ was significant at the $p < 0.006$ significance level. Like the previously discussed local polarity indices proposed by Buydens *et al.* [6], the parameter Δ differentiates solute polarities better than dipole moment.

In the same paper [10] Ośmiałowski *et al.* obtained a highly predictive QSRR describing GC retention of amines by adding energy of the highest occupied molecular orbital, E_{HOMO}, as the third parameter to regression equation:

$$I_{OV\text{-}101} = -3143.6(\pm 2284.4) - 17.38(\pm 4.15)E_T - 1920.0(\pm 875.7)\Delta - 7106.9(\pm 4856.4)E_{HOMO}; \quad (9.1)$$

$$n = 18; \quad R = 0.95; \quad s = 55.0; \quad p \leq 10^{-4};$$

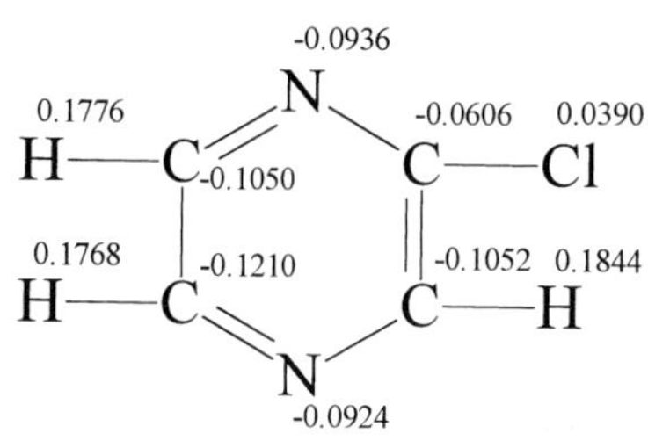

$$\Delta = 0.1844 - (-0.1210) = 0.3054$$

Figure 9.2. Electron charge distribution and determination of submolecular polarity parameter Δ for chloropyrazine.

where $I_{OV\text{-}101}$ is Kováts index determined on the methylsilicone (OV-101) stationary phase; E_T is total energy in atomic units, Δ is the polarity parameter in electrons and E_{HOMO} is the frontier orbital energy in atomic units; numbers in parentheses mean 95% confidence limits calculated by the *t*-test. Other symbols are as explained before (see for example explanations of Eqs. (8.35) and (8.36)).

Equation (9.1) may be interpreted that GC retention on a nonpolar stationary phase OV-101 increases with absolute value of total energy and energy of HOMO (negative quantities) and decreases with solute polarity Δ (positive quantity). E_T characterizes attractive dispersive interactions of solutes with stationary phase. The presence of E_{HOMO} in Eq. (9.1) could suggest that in the process of separation the electron pair donor (EPD)–electron pair acceptor (EPA) interactions take place between a donor solute molecule and the stationary phase acting as an electron acceptor. Parameter Δ in Eq. (9.1) may account for the repulsive electrostatic interactions between a polar solute and nonpolar stationary phase.

The solute polarity descriptor Δ has been found useful in QSRR studies of several sets of GC and HPLC retention data. In combination with solute bulkiness descriptors it gave a good prediction of RP HPLC capacity factors of a diverse set of benzene derivatives [11,12].

Interesting seem to be QSRR derived by Bassler *et al.* [13]. These authors analysed their HPLC capacity factors for a set of substituted benzene derivatives (including a subset of substituted phenols) determined on porous graphitic carbon stationary phase and on palladium spheres stationary phase with pure heptane as eluent.

On porous graphitic carbon the following QSRR equations were derived for substituted benzenes Eq. (9.2) and for phenolic compounds Eq. (9.3):

$$\log k' = -3.49 + 1.57\Delta - 0.277E_{HOMO}; \qquad (9.2)$$
$$n = 11; \quad R = 0.972; \quad p \le 10^{-5};$$

$$\log k' = 17.9 - 39.4\Delta - 2.45E_{LUMO}; \qquad (9.3)$$
$$n = 10; \quad R = 0.907; \quad p \le 0.003.$$

On palladium spheres the QSRR equation for phenols was

$$\log k' = 20.2 - 45.0\Delta - 2.75E_{LUMO}; \qquad (9.4)$$
$$n = 10; \quad R = 0.826; \quad p \le 0.02.$$

Equations (9.2)–(9.4) supply arguments that under nonpolar eluent conditions the porous graphitic carbon and palladium metal stationary phases behave primarily as electron pair acceptors for substituted aromatic solutes capable of *n*-donation. These results demonstrate usefulness of QSRR in studies of mechanisms of retention operating in specific chromatographic systems.

Even more convincing with that respect are QSRR describing RP HPLC retention on a polybutadiene-coated alumina (PBCA) stationary phase [14]. The set of solutes consisted of selected rigid and planar compounds (Fig. 9.3).

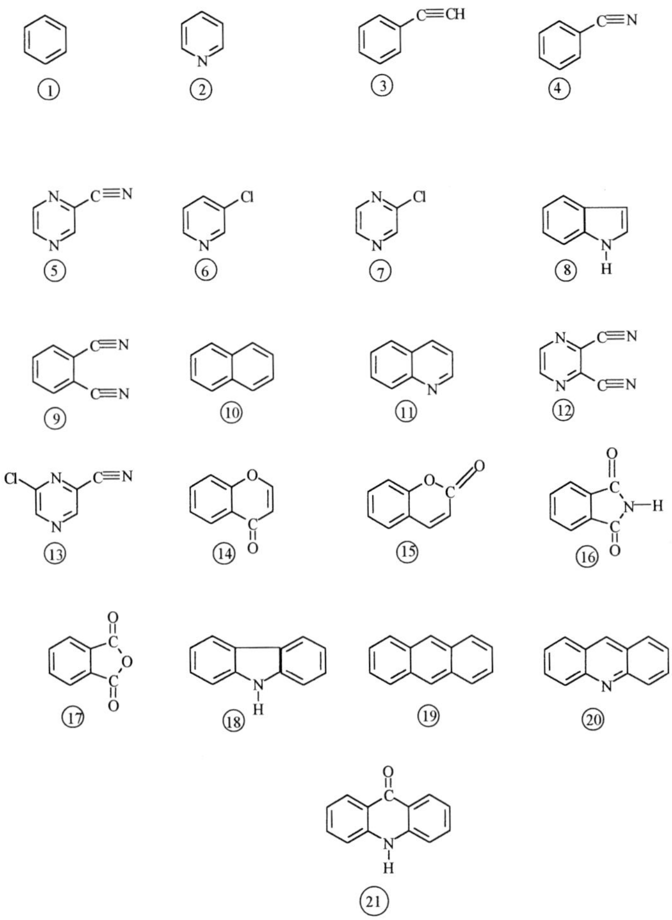

Figure 9.3. Structural formulae of test solutes chromatographed by Kaliszan and Ośmiałowski [14] on a polybutadiene-coated alumina (PBCA) RP HPLC column.

The selection of test solutes was to eliminate the possibility that the conformation of a solute interacting with the components of the chromatographic phases differed from the conformation for which structural descriptors were determined.

Due to using alumina as a support for a hydrocarbonaceous stationary phase, the PBCA columns can be operated also at alkaline pH as opposed to the standard silica-based reversed-phase materials [15]. Thus, organic base solutes can be chromatographed in nonionised forms.

The mobile phase applied by Kaliszan and Ośmiałowski [14] consisted of methanol and buffers prepared at pH 2.6 and 11.2. The ionic strength of the buffers was adjusted to 0.2 by adding $NaNO_3$. The following methanol–buffer mixtures were used as mobile phases: 80:20, 50:50, 35:65, 20:80 and 0:100 (v/v). The $\log k'$ data were plotted against the volume fraction of methanol in the eluent. The linear part of the graph was extrapolated to zero content of methanol yielding the value of $\log k'_w$ for a nonionised solute.

Sixteen various size-related, molecular graph-derived and quantum chemical (CNDO/2) descriptors were considered in QSRR analysis. The most meaningful regression equation found was

$$\log k' = -1.618 + 0.089\,\mathrm{MR} - 2.505\Delta; \tag{9.5}$$
$$n = 21; \quad R = 0.909; \quad s = 0.500; \quad F = 43;$$

where MR denotes molar refractivity calculated from bond increments [16] and F is the value of the statistical F-test; other symbols are as explained earlier. The term MR in Eq. (9.5) is significant at the level $p \leq 0.001$ and the term Δ at the level $p \leq 0.002$.

An equation of similar statistical value was obtained when using the first and second-order valence connectivity indices instead of MR. No statistically valid improvement of Eq. (9.5) was obtained by introducing additional structural descriptors considered. Only when an indicator variable, I, was introduced did the quality of correlation increase (Eq. (5.6)):

$$\log k'_w = -1.272 + 0.089\,\mathrm{MR} - 2.648\Delta - 0.598I; \tag{9.6}$$
$$n = 21; \quad R = 0.948; \quad s = 0.394; \quad F = 50.$$

The indicator variable I was assigned a value of 1 for benzene derivatives and 0 for the heterocyclic solutes chromatographed (Fig. 9.2).

Analysing Eqs. (9.5) and (9.6) one can conclude that nonspecific, dispersive interactions of solutes with the PBCA stationary phase prevail over the analogous interactions with the mobile phase. This is indicated by the positive inputs to Eqs. (9.5) and (9.6) by the structural parameters reflecting the ability of solutes to participate in dispersive (London-type) intermolecular interactions. On the other hand, the negative increment to Eqs. (9.5) and (9.6) of Δ, reflecting the ability of solutes to participate in specific (polar) intermolecular interactions, demonstrates that polar solute-mobile phase interactions are stronger than analogous solute-stationary phase interactions. Hence the RP HPLC mechanism of retention on PBCA, as revealed by the QSRR analysis, is closely similar to that on classical octadecylsilica reversed-phase material [11,12] but basically different from that postulated for normal-phase HPLC on porous graphitic carbon (Eqs. (9.2) and (9.3)) [13,17].

A slightly modified submolecular polarity parameter, $\Delta_{1,9}$, was demonstrated [18] to be a significant regressor in QSRR describing GC retention indices, I, of a series of 114 chlorinated dibenzofuranes on a nonpolar stationary phase DB-5. Chromatographic data were taken from Hale *et al.* [19] who themselves derived a regression equation relating I to a combination of 9 indicator variables of the zero–one type which represented the possible substitution patterns. The QSRR of Hale *et al.* [19], however predictive it has been, could not be interpreted in physicochemical terms. Another QSRR for this retention data set was reported by Dunn *et al.* [20]. These authors attempted to describe retention in terms of the calculated molecular surface area, S. Recalculating the Dunn *et al.* [20] data we found that molecular mass, M_m, correlated better ($r = 0.9904$) to I than S

($r = 0.9838$). Prediction of I could be improved by combining in a QSRR equation molecular mass and the polarity parameter, $\Delta_{1,9}$:

$$I = 580.3(\pm 18.9) + 5.736(\pm 0.062)M_m - 269.4(\pm 36.79)\Delta_{1,9}; \tag{9.7}$$

$$n = 114; \quad R = 0.9935; \quad s = 29.8; \quad F = 4250;$$

where $\Delta_{1,9}$ is defined as the electron excess charge difference for atoms 1 and 9 in polychlorinated dibenzofuran molecules (Fig. 9.4).

Every term in Eq. (9.7) is significant at at least $p = 0.0001$ and the t-test values corres- ponding to the consecutive terms on right side of the equation are: 30.7, 92.1 and -7.3.

The physical meaning of the term $\Delta_{1,9}$ in Eq. (9.7) deserves analysis. Introducing that parameter Ośmiałowski and Kaliszan [18] were inspired by the observation of Hale *et al.* [19] who reported that for description of retention an indicator variable was important, which assumed the value of 1 when one of the positions 1 or 9 was chlorine substituted and 0 otherwise. Hale *et al.* [19] interpreted the decreased retention of compounds with chlorine atom in one of the two positions (1 or 9) as resulting from hydrogen bonding between Cl and H. There is, however, no possibility to form such a bond because distance between the atoms in positions 1 and 9 is too large. Taking that in mind Ośmiałowski and Kaliszan [18] assumed that it was a local dipole formed by atoms in positions 1 and 9 that affected GC retention. The parameter $\Delta_{1,9}$ was designed to provide numerical measure of such a local dipole.

Pronounced effect of $\Delta_{1,9}$ on GC retention on a nonpolar stationary phase might suggest perpendicular orientation of the planar solutes towards the surface of stationary phase. In such an orientation atoms 1 and 9 of a benzofuran system are in closest contact with the nonpolar moieties of the stationary phase. At the same time, the most polar fragment of solutes (furane oxygen) is most separated from the nonpolar stationary phase. The above interpretation of mechanism of GC retention of chlorodibenzofurans reminds the "slot model" of Wise and Sander [21,22] of retention of polyaromatic hydrocarbons in RP HPLC on alkyl-bonded silica stationary phases. According to that model stationary phase consists of a number of narrow "slots" into which the planar solute molecules can penetrate depending on their shape.

The submolecular polarity parameter defined as the maximum difference of an atomic pair excess charges was found useful in QSRR studies of amide diastereomers [23,24].

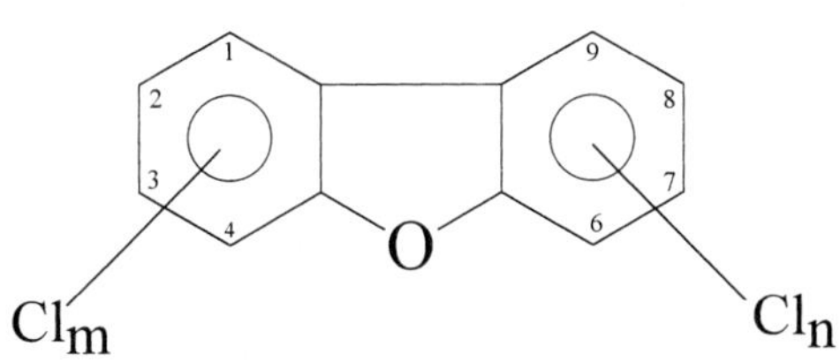

Figure 9.4. General formula and atom numbering of polychlorinated dibenzofurans. Subscripts m and n denote the number of chlorine atoms which ranges from 1 to 4.

It must be noted here that the submolecular polarity parameter Δ turned out to be of lower value for characterisation of specific interactions of solutes with polar GC stationary phases and had to be replaced in QSRR by other, however related, structural descriptors. One of such new descriptors, the topological electronic index, T^E defined in Section 4.4.8.2 (Eq. (4.20)), in combination with a dispersive interaction descriptor, total energy, E_T, gave QSRR equations predicting GC retention of a diverse set of amines on polar phases NGA and OV-225 [25].

A modification of the original Δ parameter resulted in QSRR equations describing retention of a series of benzodiazepines (Fig. 9.5) in affinity, reversed-phase and adsorption HPLC [26–28].

The modified submolecular polarity parameter, $P_{SM} = \Delta \times D_{HRN}$, was determined as follows: (i) difference in electron excess charge, Δ, between the hydrogen atom at position C_3 and the most negatively charged atom in other substituent at C_3 is

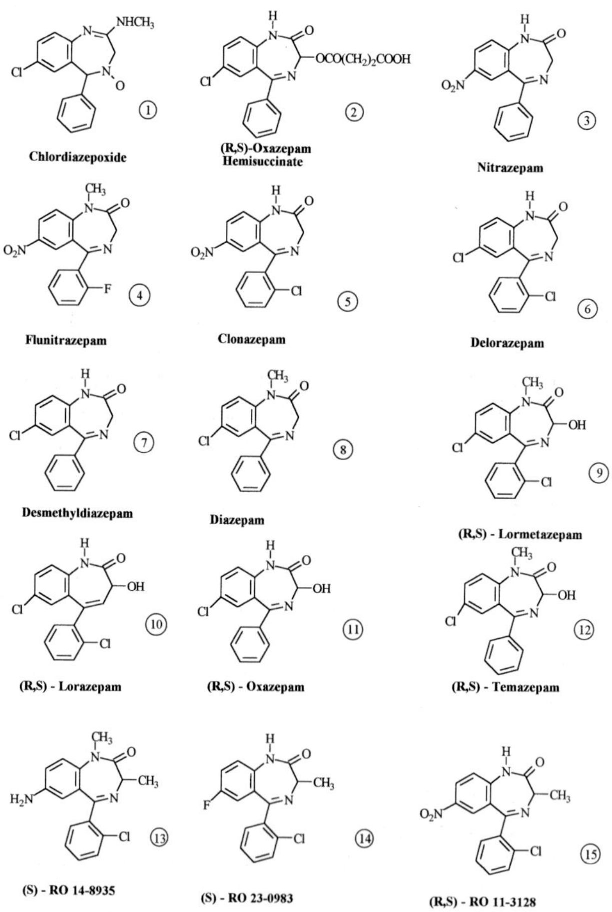

Figure 9.5.

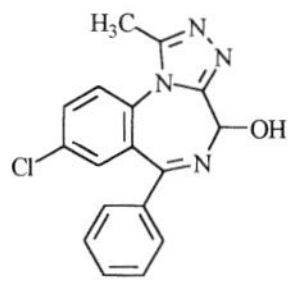

(16)

(R,S) - Alprazolam, 4-OH

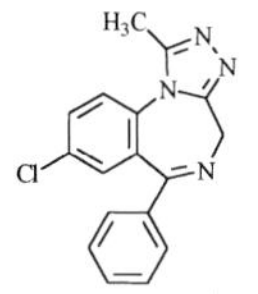

(17)

Alprazolam

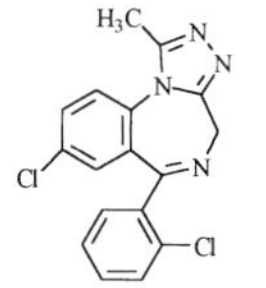

(18)

Triazolam

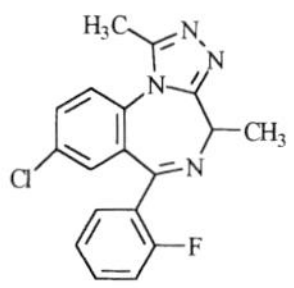

(19)

(S) - RO 11-5073

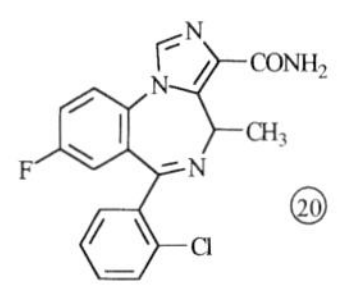

(20)

(S) - RO 23-1117

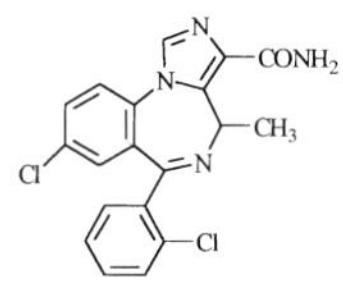

(21)

(R,S) - RO 23-3880

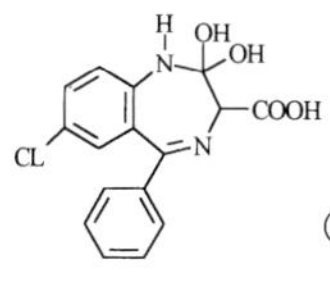

(22)

Clorazepate

Figure 9.5. Structures of 1,4-benzodiazepine test solutes.

calculated; (ii) the charge difference is multiplied by the distance D_{HRN} (in Å) between the two atoms. The excess charges on individual atoms in benzodiazepine molecule as well as interatomic distances were calculated for the geometry optimized structures. Molecular modeling was performed with the program Insight II (Biosym Technologies, San Diego, CA, USA) employing appropriate molecular orbital package (MOPAC) procedures [29,30]. Hydrophobic fragmental constants of substituents X and Y, *i.e.*, f_X, f_Y and f_{X+Y}, were according to Taylor [31]. Molar refractivity, MR, was calculated after Vogel [32].

In QSRR equations derived by Kaliszan *et al.* [26,27] employed were also the following molecular modeling derived structural descriptors (Fig. 9.6): molecular width, W, measured from the extremity of the phenyl substituent; electron excess charge on carbon atom C_3 of 1,4-diazepine system, C_3^*; energy of the lowest unoccupied molecular orbital , E_{LUMO}, and total dipole moment, μ.

The chromatographic retention parameters considered in QSRR analysis were obtained in three different HPLC systems. In an affinity HPLC model the column was packed with a human serum albumin chiral stationary phase. Chromatography was carried out isocratically. For chiral solutes $\log k'_p$ of the first eluting enantiomer and $\log k'_M$ for the second eluting enantiomer were considered separately in QSRR analysis. Separate QSRR was derived for achiral benzodiazepines.

Very good statistics characterised relationship between retention of the second eluting enantiomer (M-conformer favoured), $\log k'_M$, and of the first-eluting enantiomer

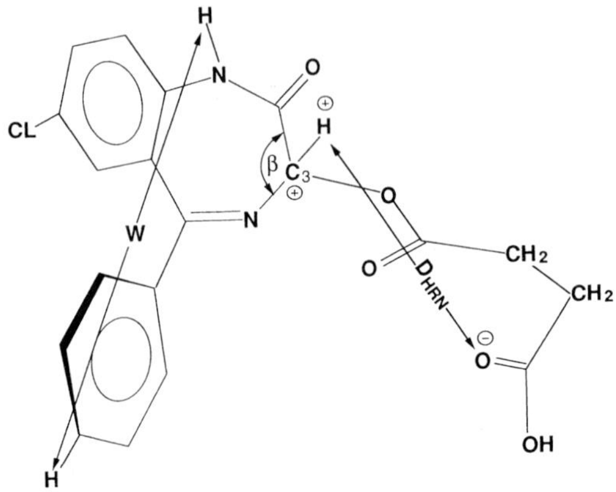

Figure 9.6. Determination of structural descriptors of benzodiazepine derivatives used in QSRR analysis. X and Y are substituents in aromatic rings; C_3^* is the excess charge on carbon C_3; W is the width of the molecule along the phenyl substituent; D_{HRN} is the distance between hydrogen at C_3 and the most negatively charged atom of another substituent at C_3. $X = H$, $Y = Cl$ in case of oxazepam hemisuccinate.

(P-conformer favoured), $\log k'_P$:

$$\log k'_M = 0.0467 + 0.8994(\pm 0.2678)\log k'_P + 0.5410(\pm 0.1167)P_{SM}; \qquad (9.8)$$
$$p \leq 1.5 \times 10^{-2} \qquad\qquad p \leq 3.6 \times 10^{-3}$$
$$n = 9; \quad R = 0.9252; \quad F = 17.8; \quad p \leq 10^{-3}.$$

All the symbols are as explained earlier. There is no intercorrelation between the independent variables $\log k'_P$ and P_{SM} ($r = 0.08$).

Within the set of HPLC parameters studied by Kaliszan *et al.* [26–28] there were 13 data points quantifying retention of the first-eluting enantiomers. The “best” QSRR for $\log k'_P$ is

$$\log k'_P = 2.4790 + 0.1834(\pm 0.0791)f_{X+Y} - 0.2779(\pm 0.0551)W; \qquad (9.9)$$
$$p \leq 0.043 \qquad\qquad p \leq 2.7 \times 10^{-3}$$
$$n = 13; \quad R = 0.8448; \quad F = 12.5; \quad p \leq 2 \times 10^{-3}.$$

The terms in Eq. (9.9) denote (see Fig. 9.6) the sum of the hydrophobicity of the substituent at position 7 in the fused benzene ring plus that of the substituent at position 2′ of the phenyl substituent, f_{X+Y}, and the width of the molecule, W. There is no intercorrelation between f_{X+Y} and W ($r = 0.17$).

With the set of eight $\log k'_M$ data points (temazepam excluded) for the second-eluting enantiomers, the stepwise regression analysis resulted in the following equation:

$$\log k'_M = 0.5558 + 0.8354(\pm 0.1540)\,P_{SM} + 0.3645(\pm 0.1987)f_{X+Y} \qquad (9.10)$$
$$p \leq 6 \times 10^{-3} \qquad\qquad p \leq 0.14$$
$$-2.6904(\pm 0.9317)C_3^*;$$
$$p \leq 4.5 \times 10^{-4}$$
$$n = 8; \quad R = 0.9384; \quad F = 9.8; \quad p \leq 0.026.$$

Intercorrelations for the pairs of variables P_{SM} *vs.* F_{X+Y}, P_{SM} *vs.* C_3^* and f_{X+Y} *vs.* C_3^* are $r = 0.25$, 0.44 and 0.14, respectively.

Regressing $\log k'_{AC}$ of achiral solutes against structural descriptors resulted in Eq. (9.11):

$$\log k'_{AC} = 1.2208 + 0.3742(\pm 0.0962)f_{X+Y} - 7.0681(\pm 1.6632)C_3^*; \qquad (9.11)$$
$$p \leq 8 \times 10^{-3} \qquad\qquad p \leq 6 \times 10^{-3}$$
$$n = 9; \quad R = 0.8893; \quad F = 11.3; \quad p \leq 0.009.$$

Equations (9.8)–(9.11) are examples of quantitative structure-enantiospecific retention relationships (QSERR). Before, there was a single report which could be classified as QSERR. Wolf *et al.* [33] succeeded in calculating capacity factors of one of the enantiomers of a series of racemate solutes chromatographed on cellulose triacetate. The authors used a negative electrostatic potential around the chiral centre of the solute, E_{00}, and the probability, Ω, of assuming a flat conformation by the molecule as variables in a regression equation. The enantiomer retention distinguishing structural parameter, P_{SM}, of Kaliszan *et al.* [26–28] is similar in nature to E_{00}. However, P_{SM} is simpler and more readily determinable. Physical meaning of Eqs. (9.8)–(9.11) will be discussed in Chapter 12 devoted to molecular recognition in chromatography.

The QSERR equations (9.8)–(9.10) concern individual enantiomers. Other reported QSERR concern rather the enantioselectivity factors, α. Norinder and Hermansson [34] attempted to relate the structure of a series of *N*-aminoalkyl-succinimides to their enantioselectivity factors observed on a chiral stationary phase based upon immobilized α_1-acid glycoprotein. To derive a model which described the behaviour of 33 solutes (of 35 solutes tested), 50 structural descriptors were subjected to a principal component analysis (PCA). The four main abstract factors extracted by PCA were then used to calculate a regression equation to predict α values. The correlation between the experimental and calculated data was characterised by $r^2 = 0.86$. To arrive at this correlation, Norinder and Hermansson [34] had to introduce 10 arbitrary indicator variables (including their squares which meaning is especially obscure). The description of enantioselectivity in terms of such abstract and arbitrary principal components obscures the interpretation of the correlation in physical terms, and, in addition, employing this approach for prediction of α for a given structure would require tedious structural assignments and recalculations.

Kaliszan *et al.* [26] were able to calculate α for their enantiomers based on individual calculated capacity factors. Thus calculated α values correlated well to the experimental ones ($r^2 = 0.9948$).

Enantioselectivity was also a subject of a QSRR study of Berthod *et al.* [35] and Carotti *et al.* [36]. These authors associated substituents with contributions to the enantioselectivity.

Prediction of retention of amide diastereomers based on principal component analysis of a variety of well designed structural descriptors was reported by Hansson and Ahnoff [23,24].

In comparative QSRR studies of retention behaviour of benzodiazepine derivatives Kaliszan *et al.* [26,27] also considered a standard reversed-phase HPLC system. The column used was packed with a specially deactivated hydrocarbonaceous silica stationary phase (Suplex pKb-100, Supelco, Bellefonte, USA). Chromatography was carried out polycratically with methanol–buffer pH 7.0 eluent. Logarithms of capacity factor, $\log k'$, for individual solutes were plotted against the volume fraction of methanol. Excellent linearity of the relationships in the whole eluent composition range studied allowed for extrapolation of $\log k'$ to 0% methanol (100% buffer). The extrapolated parameter $\log k'_w$ was described by the following QSRR equation:

$$\begin{aligned} \log k'_w = {} & 1.6541 + \underset{p \leq 10^{-5}}{0.5081(\pm 0.0422) f_Y} + \underset{p \leq 10^{-4}}{0.3084(\pm 0.0588) f_X} - \underset{p \leq 5 \times 10^{-5}}{1.6641(\pm 0.2940) C_3^*} \\ & - \underset{p \leq 2.4 \times 10^{-4}}{0.0103(\pm 0.0022)\mu^2} + \underset{p \leq 4.1 \times 10^{-4}}{0.0155(\pm 0.0034)\mathrm{MR}}; \\ & n = 21; \quad R = 0.9582; \quad F = 33.7; \quad p \leq 10^{-6}. \end{aligned} \tag{9.12}$$

Intercorrelations among individual variables are low: the highest being $r = 0.6862$ for μ^2 *vs.* MR and $r = 0.5850$ for $\log k'_w$ *vs.* f_Y.

The variables f_Y and f_X in Eq. (9.12) are highly significant statistically. The regression coefficients at these variables also differ significantly. This fact suggests the different inputs to the reversed-phase retention of benzodiazepines due to hydrophobicities of individual molecular regions.

According to Eq. (9.12), a molecular bulkiness related structural descriptor, MR, provides positive input to reversed-phase HPLC retention, whereas the molecular polarity related descriptors, C_3^* and μ^2, account for the retention decreasing effects. Kaliszan *et al.* [27] provide the following interpretation of this observation.

(i) MR reflects the ability of a solute to participate in dispersive (London-type) intermolecular interactions. These nonspecific, molecular size-dependent interactions are obviously stronger between a benzodiazepine solute molecule and a bulky hydrocarbon moiety of Suplex pKb-100 stationary phase than between the same solute molecule and small molecules forming a mobile phase. Thus, the larger MR the stronger retention.

(ii) The reverse holds true when considering parameters reflecting the ability of a solute to take part in more structurally specific, size-independent intermolecular interactions, like dipole–dipole, dipole–induced dipole, charge transfer and hydrogen bonding interactions. Such interactions, involving solute and polar molecules of eluent (water, methanol), will increase with the magnitude of polarity descriptors C_3^* and μ^2, being at the same time negligible in the case of the systems involving the solute molecule and the hydrocarbon moiety of the stationary phase.

When in QSRR analysis of the RP HPLC retention data determined on a specially deactivated hydrocarbonaceous silica stationary phase the submolecular polarity parameter P_{SM}, was considered instead of the square of total dipole moment, Kaliszan *et al.* [26,27] obtained Eq. (9.13):

$$\begin{aligned}\log k'_w = {} & 2.8824 + \underset{p \le 10^{-5}}{0.5129(\pm 0.0460)} f_Y + \underset{p \le 10^{-5}}{0.3728(\pm 0.0597)} f_X \\ & - \underset{p \le 10^{-5}}{3.0641(\pm 0.4570)} C_3^* + \underset{p \le 5 \times 10^{-4}}{0.3164(\pm 0.0719)} P_{SM}; \\ & n = 21; \quad R = 0.9478; \quad F = 35; \quad p \le 10^{-6}.\end{aligned} \tag{9.13}$$

The statistical quality of Eq. (9.13) is similar to that of Eq. (9.12). The highest intercorrelations among the variables of Eq. (9.13) are between C_3^* and P_{SM} ($r = 0.7651$) and between $\log k'_w$ and f_{X+Y} ($r = 0.7781$). The parameter P_{SM} seems to be able to replace μ^2 and MR in Eq. (9.12). This can, however, be fortuitous and discussion of P_{SM} as a net hybrid measure of dipolar and dispersive properties appears risky.

The third set of HPLC data considered in comparative QSRR studies of Kaliszan *et al.* [26–28] are capacity factors, $k'_{H\text{-}P}$ of the benzodiazepines determined on a graphitized carbon column (Shandon Hypercarb S) with hexane-propan-1-ol 97:3 (v/v) mobile phase. The QSRR equation derived for $\log k'_{H\text{-}P}$ data has a different form than the relationships derived in case of affinity HPLC (Eqs. (9.8)–(9.11)) and RP HPLC (Eqs. (9.12) and (9.13)):

$$\begin{aligned}\log k_{H\text{-}P} = {} & -1.5582 + \underset{p \le 10^{-5}}{1.0250(\pm 0.0999)} P_{SM} - \underset{p \le 10^{-5}}{1.1220(\pm 0.1564)} E_{LUMO} \\ & + \underset{p \le 0.05}{0.0078(\pm 0.0036)} \mu^2; \\ & n = 20; \quad R = 0.9428; \quad F = 43; \quad p \le 10^{-6}.\end{aligned} \tag{9.14}$$

There is an intercorrelation of $r = 0.7187$ between $\log k'_{H\text{-}P}$ and P_{SM}. Other intercorrelations are $r < 0.5$.

In Eq. (9.14) there are no solute bulkiness related structural descriptors. On the other hand, the presence of E_{LUMO}, P_{SM} and μ^2 would suggest that dispersive interactions in that normal-phase (adsorption) HPLC system are dominated by the more chemically specific directional electrostatic and electron pair donor–electron pair acceptor interactions. That observation is an agreement with the previously discussed QSRR (Eqs. (9.2)–(9.4)) describing retention of benzene derivatives on metallic stationary phases [13].

Energies of HOMO and LUMO has since long been occasionally used in QSRR studies of HPLC retention data generated on the electron pair acceptor (EPA) or electron pair donor (EPD) stationary phases. Already in 1982 Burger and Tomlinson [37] designed a simple HPLC system by dynamically coating a strong π-electron acceptor, tetracyanoethylene (TCE), on the surface of silica gel and used mixtures of hexane and 1,2-dichloroethane as eluents. A linear relationship was found between the difference of donor solute capacity factors in TCE-loaded and TCE-free columns when plotted *vs.* the EDP–EPA complex association constant, which was spectroscopically determined.

Nondek and co-workers [38–40] attempted to relate chromatographic selectivity, α_i (defined as a ratio of capacity factors for solute i and a standard), to orbital energies. The discussion of results in the original paper [40] is somewhat obscure, but it seems that bulkiness descriptors affect HPLC selectivity more than E_{HOMO} of the solute.

Profound systematic QSRR studies of TLC and HPLC retention parameters, employing quantum chemical and other calculation chemistry derived descriptors, were carried out by a British team [41,42]. Especially interesting appear regression equations describing $\log k'$ values determined on octadecylsilica columns with methanol–water, acetonitrile–water and tetrahydrofuran–water eluents (containing 1% of trifluoroacetic acid) for about 50 aromatic acids:

$$\log k'_{\mathrm{MeOH}} = 0.393\,\mathrm{CLOGP} + 0.1041\,\mathrm{DIP} + 0.114\,\mathrm{PEAX} - 0.328\,\mathrm{QON} - 1.851;$$
$$n = 46; \quad R = 0.932; \quad F = 65.7; \qquad (9.15)$$

$$\log k'_{\mathrm{ACN}} = 0.436\,\mathrm{CLOGP} + 0.119\,\mathrm{DIP} - 0.328\,\mathrm{QON} + 0.182\,\mathrm{APPDIAM} - 1.765;$$
$$n = 47; \quad R = 0.950; \quad F = 93.4; \qquad (9.16)$$

$$\log k'_{\mathrm{THF}} = 0.408\,\mathrm{CLOGP} + 0.006\,\mathrm{DIP} + 0.167\,\mathrm{HOMOE} - 4.055\,\mathrm{ESDL1} - 1.896;$$
$$n = 37; \quad R = 0.952; \quad F = 77.5. \qquad (9.17)$$

The regressors in Eqs. (9.15)–(9.17) are: CLOGP – calculated logarithm of n-octanol–water partition coefficient [43]; DIP – dipole moment [43]; PEAX principal ellipsoid axes about x [44,45]; QON – sum of the charges on the oxygens and nitrogens [43]; APPDIAM – approach diameter [43]; HOMOE – energy of the highest occupied molecular orbital [44,45]; ESDL1 – electrophilic superdelocalisability for aromatic carbon atom [45].

Equations (9.15)–(9.17) prove that calculated $\log P$ (CLOGP) and dipole moment (DIP) are most important descriptors of reserved-phase retention. Specificity of interactions of the solutes with the organic modifiers is reflected by the presence of secondary regressors in Eqs. (9.15)–(9.17). Thus, the QSRR derived are as one would expect in the case of partition liquid chromatography. The QSRR equations would be more informative from the point of view of the fundamental molecular mechanism of separation if the complex semiempirical hydrophobicity parameter (CLOGP) was replaced by simpler structural descriptors. Anyway, Eqs. (9.15)–(9.17) have some predictive potency. Even if it is limited for practical analytical purposes, yet it is realistic at the present stage of development of QSRR.

Less predictive but physically interpretable is the general equation of Cupid *et al.* [42] describing $\log k'$ data and valid for the three eluent systems:

$$\log k' = 0.501\,\mathrm{CLOGP} + 0.135\,\mathrm{DIP} + 0.109\,\mathrm{POLSOL} - 4.344; \qquad (9.18)$$

$$n = 118; \quad R = 0.909; \quad F = 177;$$

where POLSOL denotes organic modifier polarizability; CLOGP and DIP refer to solute descriptors.

Predictive value of QSRR Eqs. (9.15)–(9.18) is to low to identify a given solute peak on a chromatogram based on its structural characteristics. It is sufficiently high, however, to provide some insight into the mechanism of separation operating in the three RP HPLC systems studied. It must be noted here that predictive value of the regression equations above discussed is similar to that obtained with other chemometric approaches compared by Cupid *et al.* [42], namely the principal component analysis and neural network analysis.

Good examples of physically interpretable QSRR equations have very recently been reported by Azzaoui and Morin-Allory [46,47]. These authors studied $\log k'$ data of 36 chalcone solutes determined previously [48–50] in 18 HPLC systems using as an eluent heptane containing 0.5% of polar organic modifier. The following chemically bound silica stationary phases were employed: octadecyl (Zorbax ODS), *n*-propyl picryl ether (Spherisorb AP) and diol (LiChrospher DIOL 100). The six polar organic modifiers of the heptane-based eluent were: tetrahydrofuran (THF), dioxane (DIOX), ethanol (ETOH), propanol (PROH), octanol (OCTOH) and dimethylformamide (DMF).

In case of each stationary phase six multiple regression equations were derived (coefficients differed depending on the polar modifier in eluent) of the general form

$$\log k' = a_1\,\mathrm{Polarisability} + a_2 \log P + a_3\, E_{\mathrm{HOMO}} + a_4, \qquad (9.19)$$

where a_1–a_4 were regression coefficients. The descriptors of structure of solutes were: polarisability and energy of highest occupied molecular orbital (E_{HOMO}) calculated using the Mopac5 program [51] and logarithm of *n*-octanol-water partition coefficient, $\log P$, calculated from atomic contributions [52]. A total number of 18 equations of the form of Eq. (9.19) were derived with all the three regressors statistically significant. Significance level of individual terms of Eq. (9.19) depended on the column and polar modifier used. The authors [46,47] succeeded in deriving three more general equations valid for individual stationary phase material and all the six polar modifiers of the mobile phase. Thus, each QSRR equation accounted for 216 retention data points (36 solutes in 6 eluents). In case of Lichrospher DIOL stationary phase the equation was

$$\log k' = 0.100 \text{ Polarisability} - 0.400 \log P - 0.330\ E_{\text{HOMO}} + 1.106\ E_{\text{HOMO}}(\text{modifier}) + 0.401\ E_{\text{LUMO}}(\text{modifier}) + 6.873; \quad (9.20)$$
$$n = 216; \quad R = 0.97; \quad s = 0.097; \quad F = 655.$$

Predictive value of Eq. (9.20) is illustrated in Fig. 9.7.

QSRR equations derived by Azzaoui and Morin-Allory [47] for the two remaining stationary phases also comprised the solute polarisability and $\log P$ terms as well as the modifier E_{HOMO} and E_{LUMO}. Instead of the solute E_{HOMO} either the modifier polarisability (Spherisorb AP) or the modifier $\log P$ (Zorbax ODS) appeared statistically significant.

Analysing the regressors and regression coefficients in their QSRR equations Azzaoui and Morin-Allory [47] conclude that the solutes contributions to retention in the HPLC systems considered are similar. The main differences are due to the modifier contribution. The electron pair donor–electron pair acceptor interactions play important role in retention. These interactions are reflected by the significance in QSRR equations of the terms corresponding to energies of frontier orbitals (E_{HOMO}, E_{LUMO}).

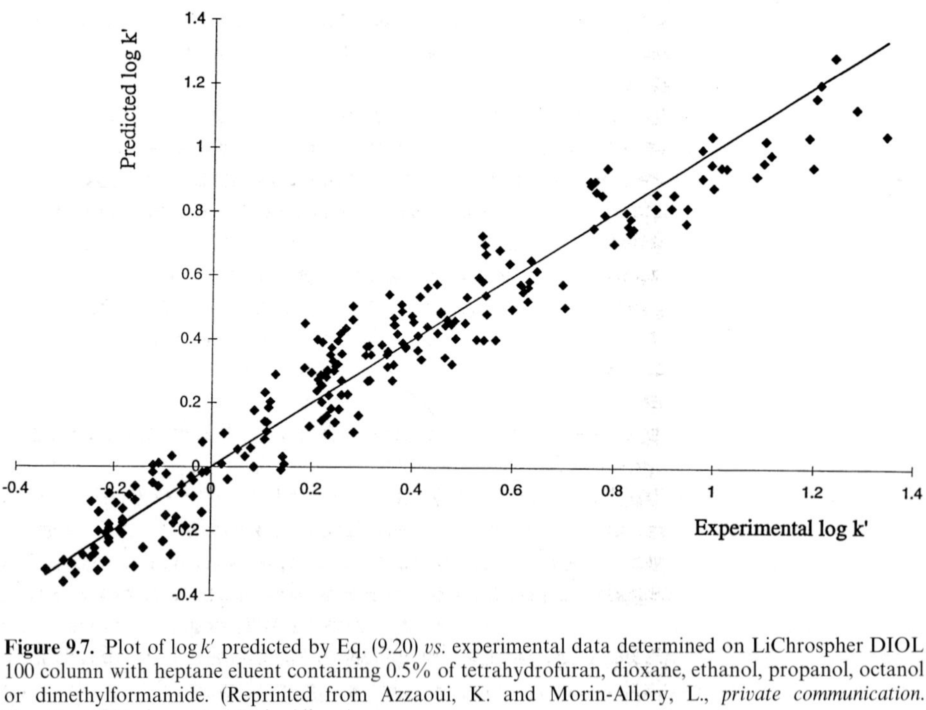

Figure 9.7. Plot of $\log k'$ predicted by Eq. (9.20) *vs.* experimental data determined on LiChrospher DIOL 100 column with heptane eluent containing 0.5% of tetrahydrofuran, dioxane, ethanol, propanol, octanol or dimethylformamide. (Reprinted from Azzaoui, K. and Morin-Allory, L., *private communication*. Courtesy of Professor L. Morin-Allory.)

The above discussed QSRR employing quantum chemical (and other) structural descriptors, have perhaps been of limited "predictive" value but shed some light onto the mechanism of chromatographic separations. There are numerous reports in the literature on QSRR aimed exclusively at retention prediction. Authors of such works looked for a set of structural descriptors (often *ad hoc* designed ones) allowing for retention predictions accurate within a few percent of the mean retention parameter value. If successful, such models could be used in lieu of authentic standards for identification of chromatographic peaks. Unfortunately, most of the models described in literature relate only to the compounds of a single class or of a few closely related classes.

Good accuracy of retention index predictions was reported in case of GC data, especially those determined on nonpolar columns. Normally, the multitude of structural descriptors considered when deriving predictive QSRR encoded, along with various quantum chemical indices, numerous topological and molecular-graph derived indices, additive bulkiness descriptors, geometrical parameters and physical properties of the solutes. Some reservations with regards to many publications are of formal statistical nature. There are intercorrelated regressors often present in the same multiple regression equation. It is often difficult to assign a definite physical sense to individual regressors. The more so if various transformations (squares, square roots, reciprocals) of descriptors are applied. The number of descriptors which can be assigned to an individual solute poses a statistical challenge. If many descriptors are tested chance correlations may result. Limited confidence in some reported predictive QSRR is also due to the fact that various equations comprise different sets of independent variables. As a matter of fact, there is no report that one researcher successfully applied for predictive purposes a QSRR equation derived by another researcher.

There is no reason, however, to discard as worthless all the predictive QSRR reported. They normally hold within the family of solutes for which they were derived and may be used for partial identification of expected chromatographic peaks. The published QSRR for given solutes chromatographed in individual systems may certainly be of help when one needs to derive predictive QSRR for similar solutes chromatographed in similar systems.

Prediction of GC retention indices based on quantum chemical indices was a subject of several publications by Massart and co-workers [6,53–56] and by Jurs and co-workers [57–60].

9.2. Prediction of Retention with the Aid of *ad hoc* Designed Descriptors

Hasan and Jurs [58] report a typical QSRR strategy for predicting retention of an unknown based on the structural features and chromatographic properties of other representative compounds. The multiparameter approach consists of generating a multitude of solute descriptors which are regressed against retention data. Observing all the statistical rules and recommendations Hasan and Jurs [58] selected the minimum number of descriptors needed to produce an equation with good predictive ability. Retention indices RI, of a series of polyhalogenated biphenyls determined by HPLC on an octadecylsilica column, with pure methanol as the mobile phase, describes the following equation:

$$\begin{aligned} \mathrm{RI} = & -6651(\pm 3646) \quad [\text{fraction of positively charged surface area}] \\ & -2469(\pm 455) \quad [\text{fraction of negatively charged surface area}] \\ & -72.9(\pm 18.8) \quad [\text{number of ortho substituents}] \\ & +3351(\pm 954) \quad [\text{relative positive charge}] \\ & -15.8(\pm 7.0) \quad [\text{path-3 kappa Kier index}]^3 + 840.2; \qquad (9.21) \\ & n = 53; \quad R = 0.968; \quad s = 55; \quad F = 285; \end{aligned}$$

where the parenthetical numbers represent 95% confidence limits. The first two descriptors in Eq. (9.21) represent the surface areas of either positively or negatively charged portions of the molecule divided by the total surface area. The relative positive charge descriptor is defined as the charge of the most positive atom in the molecule divided by the total charge of the molecule. Certainly, it is difficult to assign physical meaning to these descriptors. The least significant among the descriptors in Eq. (9.21) is a molecular graph-derived index, path-3 kappa, introduced by Kier [61] (see Section 4.4.8.2.). The path-3 kappa index might encode the "general shape of the molecules", but what would be that shape raised to the third power?

Arguing that all the published predictive QSRR have been valid only within a particular class of solutes, Katritzky *et al.* [62] proposed a method for predicting GC retention times of unknown compounds based on correlation equations derived for a large and widely diversified structurally set of organic compounds. The authors [62] determined GC retention parameters on a fused silica capillary column (cross-linked methyl silicone) for 152 randomly selected organic structures. The test series of solutes included 22 aliphatics, 89 carboxylics (59 with benzene, 10 with naphthalene and 6 with cyclohexane rings) and 41 heterocyclics (25 pyridines, 4 oxygen heterocyclics, 3 sulphur heterocyclics, 2 with piperazine and 2 with quinoline rings). Regarding the functionalities the solutes represent 31 hydroxy compounds (18 alcohols and 13 phenols), 14 ketones, 14 amines (11 primary and 3 secondary), 7 esters, 7 halogenides, 6 ethers, 6 carboxylic acids, 5 sulphides, 5 nitrides, 4 aldehydes, 2 nitro compounds and 1 sulphoxide.

The starting set of structural descriptors considered in regression analysis by Katritzky *et al.* [62] comprised 24 quantum chemical descriptors (selected out of 144 initially calculated) and 13 "conventional" additive-constitutive structural parameters, like molecular mass, number of carbon atoms, number of C–H bonds, etc. The best equation predicting retention times, t_R, of 152 solutes characterised by 37 molecular descriptors is

$$\begin{aligned} t_R = & -6.912(\pm 0.746) \quad [\text{relative number of C–H bonds}] \\ & t = -9.3 \\ & -0.871(\pm 0.102) \quad [\text{total entropy of the molecule at 300 K divided by number of atoms}] \\ & t = -8.5 \end{aligned}$$

$$+0.046(\pm 0.005) \quad [\alpha \text{ polarisability}]$$
$$t = 8.4$$
$$+0.018(\pm 0.003) \quad [\text{molecular mass}]$$
$$t = 5.9$$
$$-21.55(\pm 4.02) \quad [\text{minimum valency of an H atom}]$$
$$t = 5.4$$
$$+0.929(\pm 0.218) \quad [\text{maximum atomic orbital electronic population}]$$
$$t = 4.2$$
$$+26.50(\pm 5.98); \qquad (9.22)$$
$$t = 6.9$$
$$n = 152; \quad R^2 = 0.9590; \quad s = 0.52;$$

where in parentheses are errors of the partial regression coefficients; t are the Student's t-values for the regression terms and s is standard deviation.

According to Katritzky *et al.* [62] the most important factors in the six-parameter correlation are quantum chemical indices: α polarisability of the molecule and the minimum valency of any H atom in it. The α polarizability characterises intermolecular inductive and dispersive interactions of a solute with chromatographic medium. The minimum valency of an H atom characterises the compound as a hydrogen bonding donor. The negative value of the respective regression coefficient in Eq. (9.22) is because compounds with a lower value for minimum valency have stronger hydrogen bonds with GC stationary phase and, correspondingly, longer t_R. The negative coefficient in Eq. (9.22) at the relative number of C–H bonds the authors [62] try to explain in terms of "hydrophobic/hydrophilic interaction between the molecule studied and the GC medium". However, there is no water in the system and thus no hydrophobic/hydrophilic interactions. Negative coefficient at the relative number of C–H bonds probably reflects weaker dispersive interactions of hydrogen-substituted compounds with regards to other substituents. Positive input of molecular mass to retention is as expected. The authors [62] did not provide any explanation regarding physical meaning of the two remaining descriptors in Eq. (9.22), *i.e.*, the total entropy of the molecule at 300 K divided by the number of atoms in the molecule and maximum atomic orbital electronic population. It is difficult to guess the physical sense (if any) behind these descriptors.

There is no doubt that Eq. (9.22) provides good correlation between the experimental and calculated retention parameters. Its predictive value, however, is rather limited as evidenced by the high standard deviation (illustrated in the original paper by the scattering of points on the plot of the observed *vs.* calculated data). High correlation was obtained because a set of solutes differing more than 10 times in retention parameter was selected. To base identification of a solute peak on a chromatogram the general QSRR approach proposed by Kritzky *et al.* [62] is of too low precision. Practical applicability of predictive QSRR will for the time being remain limited to more congeneric sets of solutes. For related solutes there are

QSRR published which are better than Eq. (9.22) in terms of predictions and physical meaning [11,12,46,47,55,56].

Within series of closely related solutes even single molecular modeling-derived descriptors allow for reliable retention predictions. Not often, however, the advantages of such descriptors over simple additive structural parameters are evident [63–66]. In ref. [66] the intercept of the linear function of $\log k'$ *vs.* fraction of acetonitrile in eluent is described in terms of the water-accessible nonpolar surface area of triazine derivatives for energy minimized structures ($r = 0.942$). The slope of the same function was poorly correlated ($r = 0.878$) with the difference between the nonpolar surface area minus water-accessible nonpolar surface area. In fact, the authors [66] gave a two-parameter equation using both parameters independently (instead of their difference), but the high intercorrelation between them ($r = 0.99$) invalidates that regression equation.

Chemometrically extracted information from a variety of quantum chemical and molecular mechanics descriptors have been used for prediction of retention. Principal component regression (PCR) or partial least squares (PLS) methods were applied to correlate retention parameters with a few abstract factors concentrating the systematic information initially dispersed over a multitude of descriptors. Such obtained retention prediction has usually been more precise than that provided by regular multiple regression using a few selected descriptors [17,23,24,42]. Unfortunately, the PCR and PLS models are less convenient for practical application than multiple regression QSRR models.

There are examples of QSRR when simple structural variables accounting for the number and position of individual atoms performed not worse than more sophisticated parameters. Seybold and Bertrand [67] reported the following equation describing HPLC retention indices, RI, of polyhalogenated biphenyls:

$$\begin{aligned} \mathrm{RI} = {} & 381(\pm 16) + 144(\pm 8)N_{\mathrm{Cl}} + 171(\pm 7)N_{\mathrm{Br}} + 270(\pm 47)N_{\mathrm{I}} \\ & - 49(\pm 13)N_{\mathrm{OrF}} - 86(\pm 17)N_{\mathrm{OrCl}} - 61(\pm 20)N_{\mathrm{OrBr}} - 168(\pm 76)N_{\mathrm{OrI}}; \quad (9.23) \\ & n = 54; \quad R^2 = 0.959; \quad s = 62; \quad F = 154; \end{aligned}$$

where N_X is the number of halogens of type X and $N_{\mathrm{Or}X}$ is the number of these halogens located at the ortho position.

Equation (9.23) is of approximately the same statistical quality as the much more complex equation derived for the same HPLC data set by Hasan and Jurs [58] and discussed earlier (Eq. (9.21)).

Very good predictive QSRR regarding selectivities and response factors in GC on a 50% phenyl-50% methylpolysiloxane capillary column, were described for 220 triacylglycerols [68]. A four-parameter regression equation described selectivity values (with correlation $R^2 = 0.9993$) in terms of the total carbon number, the number of monounsaturated fatty acid residues, the number of diunsaturated fatty acid residues and the number of triunsaturated fatty acid residues.

Good predictions of GC response factors of 100 substituted benzenes and pyridines was reported based on a PLS treatment of 17 simple explanatory variables such as the number of atoms of each element, of multiple bonds, of functional groups, *etc.* [69].

9.3. Multiparameter QSRR Involving Molecular Graph Descriptors

There are numerous reports on multiparameter QSRR involving molecular graph-derived structural descriptors as regressors. In 1980 Gassiot-Matas and Firpo-Pamies [70] separated GC retention index into two components: one depending on the bulk properties of the solute and the other resulting from the polar interactions of the solute with the stationary phase. To parametrize bulkiness or steric properties of solutes, the authors [70] applied the first-order molecular connectivity index ${}^1\chi$. To characterise the polar term in correlation equations, the square dipole moment of the molecule, μ^2, was used. The general equation analysed statistically was

$$I = a{}^1\chi + b\mu^2 + c, \tag{9.24}$$

where I was the Kováts index and a, b and c were regression coefficients.

It is not clear from data of Gassiot-Matas and Firpo-Pamies [70] what is the statistical significance of the μ^2 term in their individual regression equations. For example, one set of solutes form 8 homologous hydrocarbons plus 9 aliphatic alcohols. For such a group, dipole moment, μ, assumes practically only two values, zero for hydrocarbons and about 1.7D for alcohols. In other words, the parameter μ may play the role of indicator variable accommodating two groups of solutes by one regression equation. Retention indices of each of the two groups considered separately would be well described in terms of ${}^1\chi$ by a single-parameter equation. Using any other indicator variable Ind (*e.g.*, $Ind = 1$ for hydrocarbons and $Ind = 0$ for alcohols), one would probably get the regression equation $I = f({}^1\chi, Ind)$ of the same quality as $I = f({}^1\chi, \mu^2)$.

Usefulness of molecular connectivity indices in multiparameter QSRR as the descriptors of nonspecific, dispersive increments to retention illustrate regression equations reported by Buydens *et al.* [6,7,53,54]. These authors analysed GC retention indices determined on phases of different polarity for a large set of nonheterocyclic solutes [6,7,53] and several groups of neuroleptic drugs [54]. Multiparameter QSRR equations were obtained in which various connectivity indices were present along with other topological indices, substituent electronic and hydrophobic constants and quantum chemical indices. In the case of nonpolar GC stationary phases, when the dispersion forces predominate in the intermolecular interactions, the chromatographic behaviour of solutes can be explained by means of topological indices. However, in the case of solutes of varying polarity, the significance of topological indices for prediction of retention diminishes due to differences in polar interactions with the stationary phase not accounted for by molecular connectivity. With the polar phases and structurally diverse sets of solutes, the connectivity indices become of secondary importance for retention as the specific polar interactions become decisive.

Applicability of the valence molecular connectivity indices, along with a sub-molecular polarity parameter, in regression equations describing GC and RP HPLC retention was reported by Kaliszan and co-workers [5,14]. However, molar refractivity was found to perform better in the corresponding regression equations.

There are literature reports on multiparameter QSRR in which various connectivity indices are combined with other molecular bulkiness-related descriptors like the path-2-kappa index [58], Wiener number [53], shadow areas of three-dimensional molecular models projected into three mutually perpendicular planes [58,71] and Van der Waals volume of the molecule [72,73]. Such QSRR may be useful for prediction of retention of closely related analogues but are difficult to interpret in physical terms.

The bulkiness-reflecting character of the classical first-order connectivity index, $^1\chi$ is clearly seen in a two-parameter regression equation [74] describing GC retention indices, I_N, of a series of polyaromatic hydrocarbons (PAH) on a nematic stationary phase:

$$I_N = 860\,^1\chi + 310\eta - 1710; \tag{9.25}$$

$$n = 15; \quad R = 0.995; \quad s = 109;$$

where η is a shape parameter defined as the ratio of the longer to the shorter side of a rectangle having a minimum area, which can envelop the formula of a molecule (see Section 4.4.8.1). The same type of QSRR as represented by Eq. (9.25) has been confirmed for another set of nonpolar solutes chromatographed on a GC liquid crystalline phase [75] and for several groups of nonpolar solutes chromatographed in RP HPLC systems [76,77]. Similar were the relationships reported by Yan and Martire [78,79].

Solutes considered in references [74–79] are nonpolar and planar. If a subset of isomers is selected the solutes are then of the same size (bulkiness) and η remains the only descriptor of differences in retention [76]. This observation supports the view that connectivity indices play in QSRR a role of descriptors of the ability of a solute to participate in dispersive interactions.

The question arises if these dispersive structural descriptors are the most reliable ones. Certainly, they are easily calculable and very convenient. A comparative study [18] of applications of the most commonly used structural indices derived from molecular graphs did not show their advantages in QSRR over the standard molecular size parameters.

The original Randić connectivity index provided prediction of GC retention indices of alkanes well accounting for differences in their branching [80]. This observation inclined the QSRR students to test graph theoretical indices and their transformations in various combinations from the point of view of retention prediction.

Kier and Hall [81] succeeded in relating GC retention indices of separate groups of aliphatic alcohols, ketones, esters and ethers to various connectivity indices. Correlations were good if the GC data considered were obtained on a nonpolar stationary phase. To get significant QSRR describing retention of a individual families of solutes on a polar stationary phase, different connectivity indices were used in the two-parameter equations. Different-order molecular connectivity indices and valence connectivity indices, used in various combinations, produced QSRR concerning diverse groups of solutes comprising aldehydes, ketones, esters and alcohols as well as a group of drugs including amphetamines, local anaesthetics and morphine-like compounds [82,83].

Molecular connectivity indices correlate well with retention data of related solutes. However, if the solutes differ in specific polar properties, the graph theoretical indices fail to account for such differences (even if the solutes considered are apparently similar). This was demonstrated for GC data of fatty acid methyl esters of different degree of unsaturation [84] and of aliphatic solutes comprising various functional groups chromatographed on a polar stationary phase [53]. Also, in RP HPLC, there was a lack of correlation between $\log k'$ and molecular connectivity parameters for a set of benzene derivatives containing various substituents [85]. Rather moderate, at best, have been the correlations between $\log k'$ from RP HPLC and molecular connectivity indices of a set of 12 *p*-benzoquinones and *p*-hydroquinones reported by Huang *et al.* [86]. The retention data determined on three alkylsilica phases with three different eluents yielded three-parameter correlation equations with ${}^0\chi^v$, ${}^1\chi^v$ and ${}^2\chi^v$ as the independent variables characterised by an R coefficient ranging from 0.566 to 0.877.

On the other hand there are numerous reports in which satisfactory description of GC, HPLC and TLC retention data in terms of two or more molecular graph derived indices have been reported. Earlier examples concern pyrimidine and quinazoline derivatives [87], alkylbenzenes and ketones [88,89], nitrated polyaromatic hydrocarbons [90] alkanes and their oxygenated products [91], ketones [92], barbiturates [93–99], chlorinated benzenes [100] and polyaromatic hydrocarbons [101]. Unfortunately, most of the reported correlations are fortuitous because the indices employed in multiple regression are very strongly intercorrelated. The intercorrelations were detected by Doherty *et al.* [90] and it appeared after that that no multivariant regression analysis gave statistically significantly better correlations than a single-variant analysis. Besides, questionable appear, from the point of view of physical meaning, transformations of graph theoretical indices (*e.g.*, squares, reciprocals) applied in order to get high correlations.

Even after researchers have started checking the intercorrelations among graph theoretical indices used simultaneously in individual regression equations they were quite tolerant. Several higher-order connectivity indices were employed in QSRR studies of GC data by Robbat and co-workers [102,103]. Example regression equation may be the one describing relative retention times, *RRT*, of polychlorinated biphenyls:

$$\begin{aligned} \mathrm{RRT} = & -3.5066 + 0.2455\,{}^4\chi_{\mathrm{PC}} - 0.0807({}^2\chi^v)({}^5\chi) + 0.5189\,{}^2\chi^v \\ & - 5.6936\,{}^5\chi^v_{\mathrm{C}} + 3.1571\,{}^5\chi^v; \\ & n = 206; \quad R^2 = 0.9987; \quad s = 0.0064; \quad F = 31057; \end{aligned} \tag{9.26}$$

where superscript digit denotes index order; v means valence connectivity; C (cluster) and PC (path-cluster) indicate specific methods of index calculation. The indices included in Eq. (9.26) were selected of 24 indices tested. The authors introduced "independent" variables in Eq. (9.26) providing that they were only "moderately" correlated with each other. Under "moderate correlation" the authors [103] understood Pearson correlation ≤ 0.96.

To somehow avoid the problem of intercorrelation of connectivity indices De Voogt *et al.* [104,105] attempted to introduce the $\Delta^1\chi^v$ index into QSRR describing retention of heterocyclic aromatic hydrocarbons. This index denotes the difference between the first-order valence connectivity of the heterocyclic derivatives and their parent (hydrocarbon) compounds. Two parameter regression equations comprising $\Delta^1\chi^v$ and $\Delta^1\chi^v$ terms appeared in a few instances a statistically significant improvement to simple regression. In more recent papers De Vooght and co-workers resigned of their $\Delta^1\chi^v$ parameter and reported only one-parameter regression equations predicting retention within congeneric sets of solutes [106].

Nabivach and co-workers [107,108] report using the sums of individual connectivity indices of various type and order in correlation analysis with GC retention indices of alkylphenols. It is problematic, however, if any of connectivity indices significantly improves QSRR describing retention in terms of boiling point and Van der Waals volume.

Bazylak [109] derived several multiple regression equations involving various topological indices and describing reversed-phase TLC retention of a series of 12 diethanolamines. Intercorrelation matrix provided by the author [109] clearly demonstrates that the descriptors used in the same equation are mutually correlated. Besides, introduction of individual terms does not improve the correlation significantly.

Kupchik [110,111] correctly described logarithms of GC retention times, log *T*, of a series of *n*-alkylsilanes and *n*-alkylgermanes determined on nonpolar columns by two-parameter regression equations comprising various connectivity indices. For example:

$$\log T = -0.468(\pm 0.022) + 0.485(\pm 0.004)^1\chi + 0.023(\pm 0.001)^3\chi^v; \qquad (9.27)$$

$$n = 52; \quad R = 0.999; \quad s = 0.033; \quad F = 9810;$$

where the first-order connectivity indices, $^1\chi$, and the third-order valence connectivity indices, $^3\chi^v$, are not intercorrelated significantly ($r = 0.420$). The predictive value of Eq. (9.27) cannot be questioned although it certainly is limited to the closely congeneric solutes considered.

There are several other papers in which highly predictive QSRR are reported although the intercorrelations, the multitude of indices tested and too many regressors with respect to a limited number of data points, all this makes some of the published equations disputable [112–116]. Less controversial certainly are the QSRR in which topological indices are accompanied by other, more structure specific descriptors [117] providing that not too many physically meaningless independent variables are used. The latter objection concerns a recent paper [118] in which for example a sum of χ and χ^2 multiplied by the shape parameter η is one of five independent variables used in multiparameter regression equation. Problem of nonorthogonality of molecular graph-derived descriptors has been addressed by Randić [119,120].

One approach allowing for extraction of systematic information from intercorrelated data sets is to process these data by factorial methods of data analysis. Principal component analysis (PCA) was applied to a set of molecular-graph derived and other structural parameters. Next, the first principal components were correlated by a multiparameter regression to chromatographic data. Good predictions of retention by such obtained abstract variables were reported for RP HPLC data of a set of 21 noncongeneric solutes chromatographed on a polybutadiene-coated alumina column [14]. PBCA has shown that connectivity indices, Wiener index and Kier kappa index provided the same type of structural information like the classical bulkiness parameters, *i.e.*, molecular mass, molar refractivity, *etc.* On the other hand, the information content indices (see Section 4.4.8.2.) could be classified to a group of polarity descriptors like dipole moment. Prediction of RP HPLC retention by means of PCA of topological indices was reported for barbituric acid derivatives by Cserháti *et al.* [121] and by Bazylak [122].

Regarding the molecular graph-derived indices one can argue that good retention prediction proves the validity of the index and one should try to discover the physical sense hidden in an effective descriptor. Certainly, one cannot exclude the possibility that the trial and error approach will result in the identification of universal descriptors that can easily be computed from structural formulae. The danger is that, in striving for that modern philosopher's stone, one occasionally tends to play a number game.

In recent years a few reports appeared on prediction of TLC [41], GC [123] and HPLC [42] retention data by neural networks. The application of neural networks in chromatography is still a relatively new area of research and few rules have been derived on how to build optimal networks. The published results allow for general conclusion that predictions provided by neural networks are of similar reliability as those obtained from regression models.

References

1. Streuli, C.A., Müller, W.H. and Orloff, M., *J. Chromatogr.* **101**, 17 (1974).
2. Streuli, C.A. and Orloff, M., *J. Chromatogr.* **107**, 23 (1974).
3. Gassiot, M., Fernandez, E., Firpo, G., Carbó, R. and Martin, M., *J. Chromatogr.* **108**, 337 (1975).
4. Firpo, G., Gassiot, M., Martin, M., Carbó, R., Guardino, X. and Albaiges, J., *J. Chromatogr.* **117**, 105 (1976).
5. Kaliszan, R. and Höltje, H.-D., *J. Chromatogr.* **234**, 303 (1982).
6. Buydens, L., Massart, D.L. and Geerlings, P., *Anal. Chem.* **55**, 738 (1983).
7. Buydens, L., Coomans, D., Vanbelle, M., Massart, D.L. and Vanden Driessche, R., *J. Pharm. Sci.* **72**, 1327 (1983).
8. Garcia-Raso, A., Saura-Calixto, F. and Raso, M.A., *J. Chromatogr.* **302**, 107 (1984).
9. Dubois, J.-E., Chretien, J.R., Soják, L. and Rijks, J.A., *J. Chromatogr.* **194**, 121 (1980).
10. Ośmiałowski, K., Halkiewicz, J., Radecki, A. and Kaliszan, R., *J. Chromatogr.* **346**, 53 (1985).
11. Kaliszan, R., Ośmiałowski, K., Tomellini, S.A., Hsu, S.-H., Fazio, S.D. and Hartwick, R.A., *Chromatographia* **20**, 705 (1985).
12. Kaliszan, R., Ośmiałowski, K., Tomellini, S.A., Hsu, S.-H., Fazio, S.D. and Hartwick, R.A., *J. Chromatogr.* **352**, 141 (1986).
13. Bassler, B.J., Kaliszan, R. and Hartwick, R.A., *J. Chromatogr.* **461**, 139 (1989).
14. Kaliszan, R. and Ośmiałowski, K., *J. Chromatogr.* **506**, 3 (1990).
15. Bien-Vogelsang, U., Deege, A., Figge, H., Kohler, J. and Schomburg, G., *Chromatographia* **19**, 170 (1984).
16. Vogel, A.I., Cresswell, W.T., Jeffery, G.H. and Leicester, J., *J. Chem. Soc.* **1952**, 514 (1952).
17. Kaliszan, R., Ośmiałowski, K., Bassler, B.J. and Hartwick, R.A., *J. Chromatogr.* **499**, 333 (1990).
18. Ośmiałowski, K. and Kaliszan, R., *Quant. Struct.-Act. Relat.* **10**, 125 (1991).
19. Hale, M.D., Hileman, F.D., Mazer, T., Shell, T.L., Noble, R.W. and Brooks, J.J., *Anal. Chem.* **57**, 640 (1985).
20. Dunn, W.J., Koehler, M., Stalling, D.L. and Schwartz, T.R., *Anal. Chem.* **58**, 1835 (1986).
21. Wise, S.A. and Sander, L.C., *HRC CC. J. High Resol. Chromatogr., Chromatogr. Commun.* **8**, 248 (1985).
22. Sander, L.C. and Wise, S.A., *Crit. Rev. Anal. Chem.* **18**, 299 (1987).

23. Hansson, G. and Ahnoff, M., *Chemom. Intell. Lab. Syst.* **17**, 223 (1992).
24. Hansson, G. and Ahnoff, M., *J. Chromatogr. A* **666**, 505 (1994).
25. Ośmiałowski, K., Halkiewicz, J. and Kaliszan, R., *J. Chromatogr.* **361**, 63 (1986).
26. Kaliszan, R., Noctor, T.A.G. and Wainer, J.W., *Chromatographia* **33**, 546 (1992).
27. Kaliszan, R., Kaliszan, A., Noctor, T.A.G., Purcell, W.P. and Wainer, J.W., *J. Chromatogr.* **609**, 69 (1992).
28. Kaliszan, R., Noctor, T.A.G. and Wainer, J.W., *Mol. Pharmacol.* **42**, 512 (1992).
29. Dewar, M.J.S. and Thiel, W., *J. Am. Chem. Soc.* **99**, 4899 (1977).
30. Dewar, M.J.S. and Rzepa, H.S., *J. Comput. Chem.* **4**, 158 (1983).
31. Taylor, P.J., in: "*Comprehensive Medicinal Chemistry*", Vol. 4, Hansch, C., Sammes, P.G. and Taylor, J.B. (Eds.), Pergamon Press, Oxford, 1990.
32. Vogel, A.I., "*Textbook of Practical Organic Chemistry*", Chaucer, London, 1977.
33. Wolf, R.M., Francotte, E. and Lohmann, D., *J. Chem. Soc. Perkin Trans.* 2, **1988**, 893 (1988).
34. Norinder, U. and Hermansson, J., *Chirality* **3**, 422 (1991).
35. Berthod, A., Chang, S.-C. and Armstrong, D.W., *Anal. Chem.* **64**, 395 (1992).
36. Carotti, A., Altomare, C., Cellamare, S., Monforte, A., Bettoni, G., Loiodice, F., Tangari, N. and Tortorella, V., *J. Comput.-Aided Mol. Design* **9**, 131 (1995).
37. Burger, J.J. and Tomlinson, E., *Anal. Proc.* **19**, 126 (1982).
38. Nondek, L. and Ponec, R., *J. Chromatogr.* **294**, 175 (1984).
39. Nondek, L., *J. High Resolut. Chromatogr. Chromatogr. Commun.* **8**, 302 (1985).
40. Nondek, L. and Minárik, M., *J. Chromatogr.* **324**, 261 (1985).
41. Glen, R.C., Rose, V.S., Lindon, J.C., Ruane, R.J., Wilson, I.D. and Nicholson, J.K., *J. Planar Chromatogr.* **4**, 432 (1991).
42. Cupid, B.C., Nicholson, J.K., Davis, P., Ruane, R.J., Wilson, I.D., Glen, R.C., Rose, V.S., Beddell, C.R. and Lindon, J.C., *Chromatographia* **37**, 241 (1993).
43. MEDCHEM, Version 3.52, Pomona College Medicinal Chemistry Project, Claremont, USA, 1987.
44. MOPAC, Version 5.0, Quantum Chemical Program Exchange, University of Indiana, Bloomington, Indiana, USA.
45. Glen, R.C. and Rose, V.S., *J. Mol. Graph.* **5**, 79 (1987).
46. Azzaoui, K. and Morin-Allory, L., *Chromatographia* **40**, 690 (1995).
47. Azzaoui, K. and Morin-Allory, L., *Chromatographia*, presented for publication.
48. Walczak, B., Morin-Allory, L., Chrétien, J.R., Lafosse, M. and Dreux, M., *Chemom. Intell. Lab. Syst.* **1**, 79 (1986).
49. Walczak, B., Chrétien, J.R., Dreux, M., Morin-Allory, L. and Lafosse, M., *Chemom. Intell. Lab. Syst.* **2**, 177 (1987).
50. Walczak, B., Dreux, M. and Chrétien, J.R., *Chromatographia* **31**, 575 (1991).
51. Stewart, J.J., *J. Comput.-Aided Mol. Design* **4**, 1 (1990).
52. Ghose, A.K. and Crippen, G.M., *J. Comput. Chem.* **7**, 565 (1986).
53. Buydens, L. and Massart, D.L., *Anal. Chem.* **53**, 1990 (1981).
54. Buydens, L., Massart, D.L. and Geerlings, P., *J. Chromatogr. Sci.* **23**, 304 (1985).
55. De Smet, M., Musch, G., Peeters, A., Buydens, L. and Massart, D.L., *J. Chromatogr.* **485**, 237 (1990).
56. Hindriks, R., Maris, F., Vink, F., Peeters, A., De Smet, M., Massart, D.L. and Buydens, L., *J. Chromatogr.* **485**, 255 (1990).
57. Stanton, D.T. and Jurs, P.C., *Anal. Chem.* **61**, 1328 (1989).
58. Hasan, M.N. and Jurs, P.C., *Anal. Chem.* **62**, 2318 (1990).
59. Georgakopoulos, C.G., Kiburis, J.C. and Jurs, P.C., *Anal. Chem.* **63**, 2021 (1991).
60. Georgakopoulos, C.G., Tsika, O.G., Kiburis, J.C. and Jurs, P.C., *Anal. Chem.* **63**, 2025 (1991).
61. Kier, L.B., *Med. Res. Rev.* **7**, 417 (1987).
62. Katritzky, A.R., Ignatchenko, E.S., Barcock, R.A., Lobanov, V.S. and Karelson, M., *Anal. Chem.* **66**, 1799 (1994).
63. Hanai, T., Jukurogi, A. and Huber, J., *Chromatographia* **19**, 266 (1984).
64. Yamaguchi, J. and Hanai, T., *J. Chromatogr. Sci.* **27**, 710 (1989).
65. Hanai, T., Hatano, H., Nimura, N. and Kinoshita, T., *J. Liq. Chromatogr.* **16**, 1453 (1993).
66. Valko, K. and Slégel P., *J. Chromatogr.* **592**, 59 (1992).
67. Seybold, P.G. and Bertrand, J., *Anal. Chem.* **65**, 1631 (1993).
68. Damiani, P., Sanitinelli, F., Magnarini, C., Cossignani, L. and Simonetti, M.S., *J. Chromatogr. Sci.* **32**, 21 (1994).
69. Musumarra, G., Pisano, D., Katritzky, A.R., Lapucha, A.R., Luxem, F.J., Murugan, R., Siskin, M. and Brons, G., *Tetrahedron Comput. Methodol.* **2**, 17 (1989).
70. Gassiot-Matas, M. and Firpo-Pamies, G., *J. Chromatogr.* **187**, 1 (1980).
71. Rohrbaugh, R.H. and Jurs, P.C., *Anal. Chim. Acta* **199**, 99 (1987).
72. Jinno, K. and Kawasaki, K., *Chromatographia* **17**, 337 (1983).

73. Jalali-Heravi, M. and Garkani-Nejad, Z., *J. Chromatogr.* **648**, 389 (1993).
74. Radecki, A., Lamparczyk, H. and Kaliszan, R., *Chromatographia* **12**, 595 (1979).
75. Lamparczyk, H., Kaliszan, R. and Radecki, A., *J. Chromatogr.* **361**, 442 (1986).
76. Wise, S.A., Bonnett, W.J., Guenther, F.R. and May, W.E., *J. Chromatogr. Sci.* **19**, 457 (1981).
77. Sander, L.C. and Wise, S.A., *J. Chromatogr. A* **656**, 335 (1993).
78. Yan, C. and Martire, D.E., *Anal. Chem.* **64**, 3505 (1992).
79. Yan, C. and Martire, D.E., *J. Phys. Chem.* **96**, 3489 (1992).
80. Randić, M., *J. Am. Chem. Soc.* **97**, 6609 (1975).
81. Kier, L.B. and Hall, L.H., *J. Pharm. Sci.* **68**, 120 (1979).
82. Millership, J.S. and Woolfson, A.D., *J. Pharm. Pharmacol.* **31**, 444P (1979).
83. Millership, J.S. and Woolfson, A.D., *J. Pharm. Pharmacol.* **32**, 610 (1980).
84. Kaliszan, R., *Chromatographia* **12**, 171 (1979).
85. Jinno, K. and Kawasaki, K., *Chromatographia* **18**, 90 (1984).
86. Huang, J.-X., Bouvier, E.S.P., Stuart, J.D., Melander, W.R. and Horváth, C., *J. Chromatogr.* **330**, 181 (1985).
87. Szasz, G., Papp, O., Vámos, J., Hankó-Novák, K. and Kier, L.B., *J. Chromatogr.* **269**, 91 (1983).
88. Shatz, V.D., Sakhartova, O.V., Brivkalne, L.A. and Belikov, V.A., *Zh. Anal. Khim.* **39**, 894 (1984).
89. Sakhartova, O.V. and Shatz, V.D., *Zh. Anal. Khim.* **39**, 1496 (1984).
90. Doherty, P.J., Hoes, R.M., Robbat, A. Jr. and White, C.M., *Anal. Chem.* **56**, 2697 (1984).
91. Zhang, S., Zhao, G. and Wan, S., *Sepu* **2**, 112 (1985).
92. Raymer, J., Wiesler, D. and Novotny, M., *J. Chromatogr.* **325**, 13 (1985).
93. Ekiert, L., Grodzińska-Zachwieja, Z. and Bojarski, J., *Chromatographia* **13**, 472 (1980).
94. Wells, M.J.M., Clark, C.R. and Patterson, R.M., *J. Chromatogr. Sci.* **19**, 573 (1981).
95. Bojarski, J. and Ekiert, L., *Chromatographia* **15**, 172 (1982).
96. Stead, A.H., Gill, R., Evans, A.T. and Moffat, A.C., *J. Chromatogr.* **234**, 277 (1982).
97. Wells, M.J.M., Clark, C.R. and Patterson, R.M., *J. Chromatogr.* **235**, 61 (1982).
98. Bojarski, J. and Ekiert, L., *J. Liq. Chromatogr.* **6**, 73 (1983).
99. Ekiert, L., Bojarski, J. and Mokrosz, J., in: "*Chromatography, the State of the Art*", Kalasz, H. and Ettre, L.S. (Eds.), Akademiai Kiado, Budapest, 1985, p. 403.
100. Sabljić, A., *J. Chromatogr.* **319**, 1 (1985).
101. Bulycheva, Z.Y., Rudenko, B.A., Dylevskaya, L.V. and Kutenev, V.P., *Zh. Anal. Khim.* **40**, 330 (1985).
102. Robbat, A. Jr., Corso, N.P., Doherty, P.J. and Marshall, D., *Anal. Chem.* **58**, 2072 (1986).
103. Robbat, A. Jr., Xyrafas, G. and Marshall, D., *Anal. Chem.* **60**, 982 (1988).
104. De Voogt, P. and Govers, H., *Chemosphere* **15**, 1467 (1986).
105. De Voogt, P., Wegener, J.W.M., Klamer, J.C., van Zijl, G.A. and Govers, H., *Biomed. Environ. Sci.* **1**, 194 (1988).
106. De Voogt, P., van Zijl, G.A., Govers, H. and Brinkman, U.A.T., *J. Planar Chromatogr.* **3**, 24 (1990).
107. Nabivach, V.M., Dmitrikov, V.P. and Zhigunova, T.A., *Vopr. Khim. Khim. Tekhnol.* **88**, 100 (1988).
108. Buryan, P., Nabivach, V.M. and Dimitrikov, V.P., *J. Chromatogr.* **509**, 3 (1990).
109. Bazylak, G., *J. Planar Chromatogr.* **5**, 275 (1992).
110. Kupchik, E.J., *J. Chromatogr.* **630**, 223 (1993).
111. Kupchik, E.J., *J. Chromatogr.* **645**, 182 (1993).
112. Mokrosz, J. and Duszyńska, B., *Quant. Struct.-Act. Relat.* **10**, 33 (1990).
113. Härmälä, P., Vuorela, H., Lehtonen, P. and Hiltunen, R., *J. Chromatogr.* **507**, 367 (1990).
114. Bošnjak, N., Michalić, Z. and Trinajstić. N., *J. Chromatogr.* **540**, 430 (1991).
115. Antón-Fos, G.M., Garcia-March, F.J., Pérez-Giménez F., Salabert-Salvador, M.T. and Cercós-del-Pozo R.A., *J. Chromatogr. A* **672**, 203 (1994).
116. Pyka, A., *J. Planar Chromatogr.* **8**, 219 (1995).
117. Sekuśak, S. and Sabljić, A., *J. Chromatogr.* **628**, 69 (1993).
118. Rodriguez Delgado, M.A., Sánchez, M.J., Gonzáles, V. and Garcia Montelongo, F., *J. Chromatogr. A* **697**, 71 (1995).
119. Randić, M., *J. Comput. Chem.* **12**, 970 (1991).
120. Randić, M., *J. Chem. Educ.* **69**, 713 (1992).
121. Cserháti, T., Bordas, B., Ekiert, L. and Bojarski, J., *J. Chromatogr.* **287**, 385 (1984).
122. Bazylak, G., *J. Planar Chromatogr.* **7**, 202 (1994).
123. Peterson, K.L., *Anal. Chem.* **64**, 379 (1992).

10 QSRR in Comparative Studies of Stationary Phases and Mechanism of Retention

10.1. Description of Gas Chromatographic Retention on Nonpolar and Polar Stationary Phases

Some predictive QSRR discussed in preceding chapters could comprehensibly be interpreted in physical terms. The nature of individual structural descriptors of analytes and their significance for the QSRR derived clearly reflected the characteristical features of separation system. By means of QSRR the differences among chromatographic systems could be quantified as regards to the prevailing mode of distribution of solutes between phases (partition, adsorption) and specific properties of stationary (mobile) phases. It is often difficult to separate "predictive" and "mechanistic" aspects of QSRR. In this chapter, however, the emphasis will be put on comparison of various chromatographic systems and not of the test solutes which are employed to derive QSRR.

Analysing the intermolecular interactions involved in the chromatographic process one can conclude that generally two main structural features of solutes affect their retention. One type would be related to molecular bulkiness and would reflect the ability of a solute to undergo interactions which are usually termed nonpolar or nonspecific. The other type of structural parameters providing an input to retention are the so-called polar, or chemically specific, structural descriptors. Hence a free-energy related retention parameter, *Ret*, could be described by a general equation [1]:

$$\mathrm{Ret} = af(P) + bf(NP) + c, \qquad (10.1)$$

where $f(P)$ is a fuction of the ability of a solute to undergo polar interactions, $f(NP)$ is a function reflecting the tendency of a solute to partcipate in molecular interactions of a nonpolar type, and a, b and c are constants being functions of properties of the chromatographic system.

Analysing GC retention indices of a large group of substituted phenols Kaliszan and Höltje [2] started with the following regression equation:

$$I = a(\mathrm{MR}) + b\mu^2 + c, \tag{10.2}$$

where I were Kováts indices determined by Grzybowski *et al.* [3] on a nonpolar phase dimethylpolysiloxane (SE-30) and two polar phases: 3-cyanopropylmethylpolysiloxane (OV-225) and polyneopentyl glycol adipate (NGA). In Eq. (10.2) MR denotes molar refractivity and μ is dipole moment; a, b and c are regression coefficients.

For 20 phenol derivatives, of total 43 chromatographed, the authors [2] found experimental dipole moments in the literature; MR data were calculated according to Hansch *et al.* [4]. Regression equations of the form of Eq. (10.2) were characterised by low correlation coefficients. Statistically significant appeared these regressions if four solutes of 20 were excluded from analysis. It was concluded that dipole moment, experimentally determined in benzene solutions, is only a crude approximation of the polar interaction contribution to the GC retention index. Thus, if the separation of chromatographic interactions into two distinctive classes, *i.e.*, nonpolar and polar interactions, is justified, a more adequate parametrization of solute polarity is required for successful QSAR.

To arrive at a gas chromatographic polarity measure the following reasoning was carried out [2,5]. Let one assume P and D to be reliable solute polar and dispersive properties descriptors, respectively. If one has retention indices for a given set of compounds determined on a polar phase, I_P, and on a nonpolar phase, I_{NP}, then one can write

$$I_P = a_P P + b_P D + c_P, \tag{10.3}$$

$$I_{NP} = a_{NP} P + b_{NP} D + c_{NP}, \tag{10.4}$$

where a_P, b_P, c_P, a_{NP}, b_{NP} and c_{NP} are constant for polar (subscript P) and nonpolar (subscript NP) phases. From Eq. (10.4) one gets for the descriptor P:

$$P = (I_{NP} - b_{NP} D - c_{NP})/a_{NP}. \tag{10.5}$$

Now Eq. (10.3) can be rewritten as

$$I_P = [a_P(I_{NP} - b_{NP} D - c_{NP})/a_{NP}] + b_P D + c_P. \tag{10.6}$$

After rearrangement, one gets

$$I_P = (a_P/a_{NP}) I_{NP} - [(a_P/a_{NP}) b_{NP} - b_P] D + \mathrm{const}. \tag{10.7}$$

Denoting new constants as k_1, k_2 and k_3, Eq. (10.7) gets the form

$$I_P = k_1 I_{NP} - k_2 D + k_3. \tag{10.8}$$

Constant k_1 equals a_P/a_{NP}, k_2 equals $(a_P b_{NP}/a_{NP} - b_P)$ and k_3 corresponds to c_P.

If the general relationship Eq. (10.1) is valid, then for the same sets of solutes, retention indices determined on polar phases should be related to the respective data determined on nonpolar phases by an equation of the form of Eq. (10.8) involving a negative term proportional to a solute dispersive properties descriptor. The greater the differences in polarities of the stationary phases, the larger is the absolute value of the k_2D term in Eq. (10.8). There would be a change of the sign at the k_2D term if the retention index on a nonpolar phase was described in terms of the index obtained on a polar phase and a dispersive descriptor.

The validity of Eq. (10.8) was demonstrated [2] based on the experimental data of Grzybowski *et al.* [3]. The following regression equations relate Kováts indices of phenols determined on polar phases OV-225 and NGA to molar refractivity, MR, and Kováts indices determined on a nonpolar phase SE-30:

$$I_{\text{OV-225}} = 1.95(\pm 0.16) I_{\text{SE-30}} - 22.08(\pm 3.56)\text{MR} + 285.76; \qquad (10.9)$$

$$n = 43; \quad R = 0.969; \quad s = 71;$$

$$I_{\text{NGA}} = 1.81(\pm 0.20) I_{\text{SE-30}} - 24.50(\pm 4.36)\,\text{MR} + 682.59; \qquad (10.10)$$

$$n = 43; \quad R = 0.947; \quad s = 87.$$

Intercorrelations between regressors are lower than $r = 0.3$.

Relationships of the form of Eqs. (10.9) and (10.10) were noted in the original report by Grzybowski *et al.* [3] but these authors could not rationalize their observation and erroneously interpreted them claiming that on polar phases retention decreases with molar refractivity.

Another evidence for validity of Eq. (10.8) can be deduced from data provided by Halkiewicz *et al.* [6]. These authors related GC retention time, *t*, on polar (NGA and OV-225) and nonpolar (OV-101) stationary phases for a series of copper and nickel complexes of dialkyldithiocarbamates with the help of the parameter *L* reflecting the length of the alkyl substituent. One of their equations is

$$t_{\text{OV-225}} = 5.10(\pm 0.60)\, t_{\text{OV-101}} - 4.53(\pm 1.49)\, L + 17.42(\pm 5.06); \qquad (10.11)$$

$$n = 15; \quad R = 0.984; \quad s = 2.38.$$

As a matter of fact Halkiewicz *et al.* [6] incorrectly regard the length of the alkyl substituent, *L*, as a measure of steric hindrance decreasing retention on polar phases. In view of the above discussed Eq. (10.8) the *L* parameter must be treated as a measure of solute bulkiness or as a dispersive property descriptor.

There is a rather lengthy argumentation [7] that QSRR describing GC retention in terms of boiling point and Van der Waals volume as reported by Bermejo *et al.* [8–10], can also be rationalised in a manner similar to that expressed by Eqs. (10.1)–(10.8). In the same reference [7] the determination of GC polarity parameter is described based on QSRR equations combining retention indices from phases of different polarities. This polarity parameter correlates only moderately to total experimental dipole moment of solutes determined in benzene solution. This observation proves a limited suitability of the total dipole moment as a descriptor of the ability of solutes to participate in polar interactions. On the other hand, molar refractivity seem to be quite an adequate descriptor of dispersive intermolecular interactions determining GC separations.

Kleinert *et al.* [11,12] in their studies on GC retention analysed the following relationship:

$$\log V_g = A_0 + A_1\Psi + A_2D, \tag{10.12}$$

where V_g is the specific retention volume and A_0–A_2 are empirical constants. The variables Ψ and D are to characterise polar and nonpolar properties of solutes, respectively. These parameters are calculated on the basis of a simplified geometric model.

According to Kleinert *et al.* [11,12], the parameter D is a measure of dispersive interactions between solute and stationary phase molecules. These authors proposed the following formula to calculate D:

$$D = \sum_{i=1}^{m} \sum_{j=1}^{n} D_{ij}K_iW_j, \tag{10.13}$$

where D_{ij} are the interaction contributions of all of the interacting groups i and j of the solute and the stationary phase molecules, respectively. These interaction contributions are summed after correction due to the shielding factor K_i and the contact point concentration W_j. The number of interacting groups i of the solute is m, and the corresponding number for the stationary phase molecule is n. The individual contributions, D_{ij}, are calculated from the interacting groups refractions $(\mathrm{MR})_j$ and the distance r_{ij} between the interacting groups (r_{ij} is the sum of the group Van der Waals radii):

$$D_{ij} = [(\mathrm{MR})_i(\mathrm{MR})_j]/r_{ij}^6. \tag{10.14}$$

The correction factors K_i and W_j in Eq. (10.13) require explanation. K_r and K_j defined as the numbers of contact points of an atom group of solute i and stationary phase molecule j, are the numbers of sites on which the molecule is able to interact with neighbouring molecules. It is assumed that each atomic group can interact with the four nearest neighbours, and consequently, the value of K_i or K_j is 4 if interacting groups are not shielded and 0 if the groups are completely shielded. The factor W_j is the concentration of contact points belonging to the atomic group j in the molecule of stationary phase. The W_j value is calculated as the ratio of all interacting contact points of atomic group j, *i.e.*, $\sum K_j$, to the total number of contact points of all different atomic groups present in the stationary phase molecule.

The parameter Ψ in Eq. (10.12) is considered by Kleinert and Ecknig [11] as a measure of the energy of intermolecular interactions due to polar functional groups. The authors carried out calculations in case of dipole–dipole orientation interactions, Ψ_{or} and in case of dipole–induced dipole interactions, Ψ_{ind}. Generally, to calculate Ψ (either Ψ_{or} or Ψ_{ind}) they employed the equation:

$$\Psi = \sum_{u=1}^{m'} \sum_{v=1}^{n'} \Psi_{ij}^{uv} K_\mu W_v, \tag{10.15}$$

where u and v are the so-called contact segments belonging to the polar atomic groups in solute i and stationary phase j. The contact segments are chosen according to the following rules: (a) each atom may have four nearest neighbours; (b) the orientation of the interacting groups, which provide minimum potential energy are preferred; (c) all of the contact segments of the solute interact with the stationary phase molecule if there is no steric hindrance to the interaction. The contact segment is understood as that part of the molecular surface where the polar groups may contact during the intermolecular interaction. The correction factor K_μ is the number of contact segments, μ, characterised by equal interaction energy. The factor W_v is defined as the ratio of the sum of energetically equal contact segments v, *i.e.*, $\sum v$, to the total sum of all contact segments of the stationary phase molecule which are geometrically able to interact with the corresponding contact segments u of the solute molecule.

If Ψ in Eq. (10.15) is to represent dipole–dipole interactions (orientation interactions) then $\Psi^{uv} = \Psi_{ij}^{\text{or}}$ is calculated from the classical Keesom formula:

$$\Psi_{ij}^{\text{or}} = \mu_i \mu_j[-2\cos\Theta_i \cos\Theta_j + \sin\Theta_i \sin\Theta_j + \cos(\Phi_i - \Phi_j)]/r_{ij}^3, \qquad (10.16)$$

where μ_i and μ_j are the dipole moments of the groups i and j of solute and stationary phase molecules, respectively; r_{ij} is distance between geometrical centres of the dipoles and Θ_i, Θ_j, Φ_i and Φ_j are orientation angles.

Regarding the induction forces Kleinert and Ecknig [11] applied the classical formula of Debye:

$$\Psi_{ij}^{\text{ind}} = \mu_i^2 \alpha_j (3\cos\Theta_i + 1)/(2r_{ij}^6), \qquad (10.17)$$

where α_j is the group polarizability of the interacting group j.

To verify the above described nonempirical approach to GC retention evaluation Kleinert and Ecknig [11] considered the specific retention volumes, V_g, determined on three phases of different polarities. The set of solutes analysed consisted of 27 aliphatic alcohols and ketones, halogenoalkanes and nitroalkanes. The dispersion descriptor D was calculated for the interaction with each of the three stationary phases separately. In the case individual solutes the value of D was lowest for the interaction with the OV-17 phase, higher for the solute-phase OV-101 interaction and the highest in the case of solute interaction with the ODPN phase. As expected, for an individual phase the parameter D increased with the chain lengths of homologues.

Kleinert and Ecknig [11] calculated the polar parameters Ψ_{ij}^{ind} or Ψ_{ij}^{or} only for interactions with the polar phases, OV-17 and ODPN. As expected, these polar parameters are the same for all members of a homologous series.

With the arduously determined variables Ψ and D Kleinert and Ecknig [11] derived QSRR equations of the form of Eq. (10.12). In the case of nonpolar OV-101 phase linear relationship $\log V_g$ *vs.* D was characterised by $r = 0.880$. For the polar phase OV-17 the two-parameter regression equation with D and Ψ^{ind} as regressors gives $R = 0.901$. An analogous QSRR describing V_g determined on the ODPN phase in terms of D and Ψ^{or} is characterised by $R = 0.899$.

Certainly, the predictive value of QSRR resulting from the Kleinert *et al.* [11,12] approach is not impressive. Of a higher predictive potency are QSRR based on structural descriptors obtained by molecular modeling and other QSRR reviewed in Chapters 7–9. Besides, the determination of D and, especially, Ψ parameters is relatively complex. However, the approach proposed by Kleinert *et al.* [11,12] clearly demonstrates fundamental molecular mechanism determining separations and rationalizes observed differences in retention due to properties of stationary phases.

10.2. Description of Retention in Reversed-Phase Liquid Chromatography

There are three main variables determining distribution of a solute between a mobile and a stationary phase in liquid chromatography. At constant temperature of separation, the three variables are: chemical structure of the solute, physicochemical properties of the mobile phase and physicochemical properties of stationary phase. If one gets numerical measures of properties of the solutes, of the mobile phases and of the stationary phases, one can derive a general relationship linking these properties to retention parameters. One such attempt was undertaken by Jinno and Kawasaki [13–15].

Jinno and Kawasaki [13–15] found that for a set of substituted benzene derivatives the $\log k'$ data, determined on four reversed-phase materials with acetonitrile-water and methanol-water eluents of different composition, can be described by regression equations of the form:

$$\log k' = g\pi + h(\mathrm{HA} - \mathrm{HD}) + i, \tag{10.18}$$

where π is the Hansch hydrophobic substituent constant (see Section 4.2), HA is the number of hydrogen acceptor groups in the solute molecule and HD is the number of hydrogen donor groups. The regression coefficients g, h and i depend on the nature of organic modifier (acetonitrile, methanol) and its volume fraction in eluent. Jinno and Kawasaki [13–15] succeeded in deriving regression equations relating each of the coefficients of Eq. (10.18), *i.e.*, g, h and i, to volume fraction X of acetonitrile, as well as methanol, in eluent. These functions are complex and difficult to rationalize: they comprise several X^n terms (where $n = 0, 1, 2, 3, 4$). Nonetheless, there is a consistency between the observed and predicted RP HPLC retention parameters for several chromatographic systems involving three distinctive classes of solutes (*i.e.*, alkylbenzenes and polyaromatic hydrocarbons, substituted benzenes and phenols), four reversed-phase columns (*i.e.*, C_{18}, C_8, C_2 and bonded phenyl) and two binary eluents of varying composition (*i.e.*, acetonitrile-water and methanol-water). A computer assisted retention prediction system for RP HPLC was proposed [14,15]. To calibrate the program, one needs some retention data for standard substances (four or five according to the authors) on every column one wants to use, at several eluent compositions.

The QSRR-based expert system developed by Jinno and Kawasaki [13–15] is very instructive as regards reversed-phase liquid chromatographic separations. Its main shortage is a limited availability of structural descriptors of solutes requested for Eq. (10.18). The basic solute descriptor applied is the hydrophobic constant π of substituents. Excepting common aromatic compounds the quantitation of hydrophobicity of solutes cannot be done by means of π and is of limited reliability if the calculation chemistry methods are applied (see Section 7.2).

The parameter π characterises partition properties of solutes. The RP HPLC is a separation method also based on solute partition between two phases. Hence, the dependence of $\log k'$ from RP HPLC on π, with a correction for hydrogen bonding (HA–HD) Eq. (10.18), could be expected. Jinno and Kawasaki [13–15] provided quantitative (statistical) proof for that.

A more general QSRR approach to studies of RP HPLC retention was reported by Kaliszan *et al.* [16,17]. The QSRR equations were derived which account for changes (limited) of all three chromatographic variables, *i.e.*, solute structure, mobile phase composition and stationary phase properties. Test series of solutes consisted of 12 benzene derivatives possessing various functional groups (Table 10.1).

The mobile phases were mixtures of methanol and water of composition ranging from 35% to 65% of methanol. The lower and the upper concentrations limited the linearity of the dependence of $\log k'_{ij}$ (for the i-th solute determined on the j-th stationary phase) on the mole fraction X of one of the components of the binary eluent (here it has been water):

$$\log k'_{ij} = a_{ij} X + b_{ij}, \tag{10.19}$$

where a_{ij} and b_{ij} are constants for a given solute i chromatographed on an individual stationary phase j.

The third variable in the RP HPLC systems studied by Kaliszan *et al.* [16,17] was the dimethyloctadecylsilane (C_{20}) coverage of chemically bonded stationary phase. Three columns of C_{20} coverage 1.54×10^{-4}, 3.18×10^{-4} and 4.96×10^{-4} mol/g were taken into consideration. The $\log k'$ values determined on a column of C_{20} coverage of 6.6×10^{-4} mol/g were not used in QSRR studies because of evident deviation from the linearity of the $\log k'$ *vs.* C_{20} coverage plots for that high coverage.

For the solutes, the stationary phases and the methanol-water composition range studied, linearity was observed ($r = 0.99$) of $\log k'$ *vs.* X plots. Thus, the coefficients a_{ij} and b_{ij} of Eq. (10.19) could be determined by linear regression. Next, the coefficients a_{ij} and b_{ij} determined for an individual solute i on the stationary phase of hydrocarbon

Table 10.1 Test solutes and their structural descriptors calculated by the CNDO/2 molecular orbital method [16,17].

No.	*Compound*	*Total Energy (Atomic Units)*	*Maximum Excess Charge Difference (Electrons)*	*Dipole Moments (D)*
1	Phenol	−65.5548	0.4328	1.749
2	Acetophenone	−81.1756	0.5088	3.042
3	Nitrobenzene	−94.8446	0.7774	5.059
4	Methylbenzene	−99.6184	0.6836	2.038
5	*p*-Cresol	−74.2375	0.4265	1.738
6	*p*-Ethylphenol	−82.6639	0.4275	2.332
7	*p*-Propylphenol	−91.3479	0.4270	2.147
8	4-sec-Butylphenol	−100.1758	0.4246	1.846
9	Aniline	−59.5473	0.3737	1.521
10	*N*-Methylaniline	−68.2305	0.3564	1.150
11	4-Chloroacetophenone	−96.6241	0.5028	2.289
12	3,4-Dichloroacetophenone	−112.3142	0.4922	1.281

coverage j were described in terms of the quantum chemically calculated total energy of a solute, E_{Ti}, and its submolecular polarity parameter Δ_i, defined as the maximum atomic excess charge difference in a molecule (see Section 4.4.6):

$$a_{ij} = \alpha_j E_{Ti} + \beta_j \Delta_i + \gamma_j, \tag{10.20}$$

$$b_{ij} = \alpha'_j E_{Ti} + \beta'_j \Delta_i + \gamma'_j, \tag{10.21}$$

where α_j, β_j, γ_j, α'_j, β'_j and γ'_j are regression coefficients. For example parameters a_{ij} and b_{ij} determined for phenol ($i = 1$) on the stationary phase having 3.18×10^{-4} mol/g C_{20} coverage ($j = 2$) were described by the following equations:

$$a_{12} = 0.5340(\pm 0.0122)E_{Ti} + 3.3473(\pm 1.5866)\Delta_i - 0.4034(\pm 0.9334); \tag{10.22}$$

$$n = 12; \quad R = 0.962; \quad s = 0.2481;$$

$$b_{12} = -0.0415(\pm 0.0068)E_{Ti} - 2.3986(\pm 0.8663)\Delta_i - 0.9429(\pm 0.5083); \tag{10.23}$$

$$n = 12; \quad R = 0.980; \quad s = 0.1351.$$

The a_{ij} and b_{ij} data for i compounds chromatographed on j stationary phases are determined from $i \times j$ regression equations of the form of Eq. (10.19). If Eq. (10.20) and (10.21) are of good statistical quality, Eq. (10.19) for the j-th phase may be rewritten as

$$\log k'_{ij} = (\alpha_j E_{Ti} + \beta_j \Delta_i + \gamma_j) X + (\alpha'_j E_{Ti} + \beta'_j \Delta_j + \gamma'_j). \tag{10.24}$$

The stationary phases j differed in their C_{20} coverage. The authors [16,17] found linearity for a given solute i chromatographed at a fixed eluent composition X, between $\log k'_{ij,X}$ and C_{20} stationary phase coverage, C_j, for three of their four phases:

$$\log k'_{ij,X} = AC_j + B, \tag{10.25}$$

where A and B are regression coefficients. In such a situation it was assumed that the coefficients a_{ij} and b_{ij} of Eq. (10.19) depend not only on the solute structure described by individual E_{Ti} and Δ_i parameters but also linearly (over a limited range) on the stationary phase C_{20} coverage, C_j. In result, the general equation describing logarithms of capacity factors of solute i chromatographed on a hydrocarbon-bonded silica stationary phase j of C_{20} coverage C_j with methanol-water eluents of the mole fraction of water X is

$$\log k'_{ij} = (\alpha E_{Ti} + \beta \Delta_i + \gamma C_j + \delta) X + (\alpha' E_{Ti} + \beta' \Delta_i + \gamma' C_j + \delta'). \tag{10.26}$$

The QSRR equation derived by Kaliszan *et al.* [16,17] was

$$\begin{aligned} \log k'_{ij} = {} & [0.0454(\pm 0.0071)E_{Ti} + 2.6493(\pm 0.9187)\Delta_i - 0.1053(\pm 0.0672)C_j \\ & - 0.4946(\pm 0.5828)]X + [-0.0381(\pm 0.0039)E_{Ti} \\ & - 2.1659(\pm 0.4919)\Delta_i + 0.1696(\pm 0.0359)C_j - 1.2963(\pm 0.3120)]. \end{aligned} \tag{10.27}$$

The statistical parameters are $n = 36$, $R = 0.925$ and $s = 0.2756$ for term a_{ij} and $n = 36$, $R = 0.972$ and $s = 0.1476$ for term b_{ij} of Eq. (10.19). The correlation between

144 experimental $\log k'_{ij}$ and the values calculated by Eq. (10.27) is $r = 0.986$. This correlation is illustrated in Fig. 10.1.

It appeared interesting to compare the predictive value of Eq. (10.27) with an analogous QSRR equation in which the submolecular polarity parameter Δ_i was replaced by the CNDO/2-calculated total dipole moment μ_i. The statistics obtained were significantly lower. For a_{ij} calculated in terms of E_{Ti}, μ_i and C_j, the correlation coefficient was $R = 0.871$; for b_{ij}, the corresponding value was $R = 0.904$. It should be noted here that the correlation between μ_i and Δ_i for 12 solutes was $r = 0.769$. This correlation is rather high to use μ_i and Δ_i in the same regression equation. For the same reason, energies of the highest occupied molecular orbital, E_{HOMO}, and the lowest unoccupied molecular orbital, E_{LUMO}, were not included in multiparameter regression equations.

As illustrated in Fig. 10.1 the prediction of RP HPLC retention in terms of solute structure, mobile phase composition and stationary phase properties is possible, even if precision might be too low for identification purposes. Again, two main structural factors determine solute retention as quantified by a bulkiness parameter, E_{Ti}, and

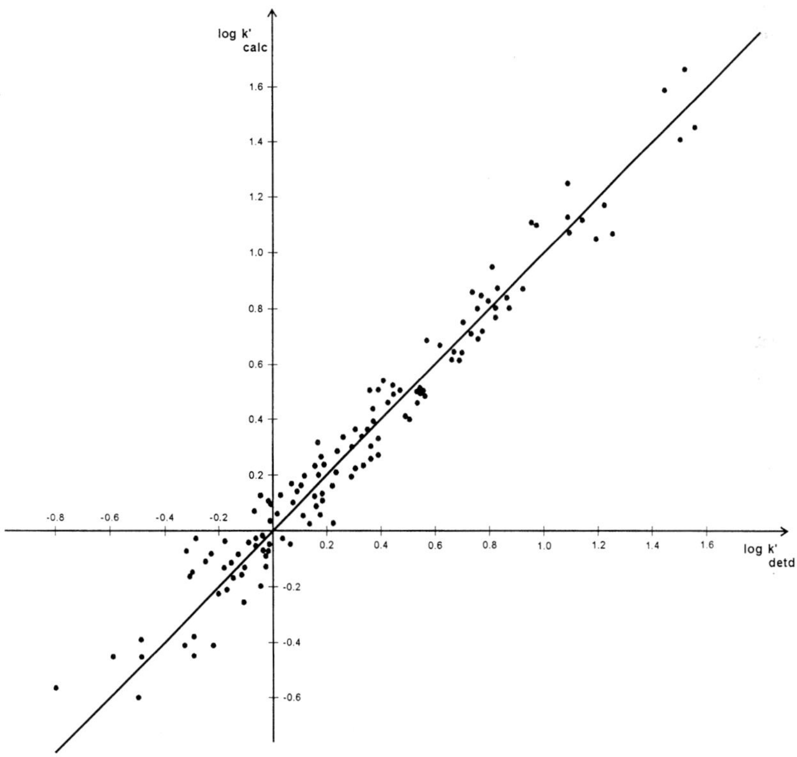

Figure 10.1. Correlation of predicted by Eq. (10.27) ($\log k'_{\mathrm{calc}}$) and observed ($\log k'_{\mathrm{detd}}$) RP HPLC retention parameters for 12 test solutes chromatographed on three stationary phases at five methanol-water eluent compositions. (Reprinted from Kaliszan, R., Ośmiałowski, K., Tomellini, S.A., Hsu, S.-H., Fazio, S.D. and Hartwick, R.A., *J. Chromatogr.* 352, 141, 1986 with kind permission of Elsevier Science-NL, Sara Burgerhartstraat 25, 1055 KV Amsterdam, The Netherlands.)

a polarity parameter, Δ_i. For the solutes selected and the chromatographic systems considered, other structural parameters appeared insignificant for retention. In other chromatographic systems, for instance, the energies of frontier orbitals, E_{HOMO} and E_{LUMO}, may appear significant for retention. One has to be cognizant of the fact that QSRR are derived statistically. Lack of significance of a given property parameter (descriptor) needs not to result from the absence of intermolecular interactions characterised by this parameter during the separation process. Formally statistically insignificant will be the properties (parameters) which provide similar contribution to retention for a great majority of test solutes. Thus, for a series of solutes of closely similar size (bulkiness) the dispersive interaction parameters will apparently be insignificant for QSRR.

When the descriptors of the abilities of solutes to participate in a given type of intermolecular interactions are not precise enough (and this is often the case), then their significance may not manifest itself in QSRR, especially if the test series of solutes is relatively short. This also happens if the interaction concerned is of secondary importance for retention in the system considered. For example, it might be difficult to prove by QSRR the importance for the HPLC retention of dispersive, polar and charge transfer interactions for a group of polychlorinated dibenzo-*p*-dioxins although the original authors [18] provide convincing explanation of the observed chromatographic behaviour in qualitative terms. The reason is that the differences in solute bulkiness, and thus in dispersive interactions, dominate other potential terms in correlation equations. At the same time, the common measures of solute polarities (*e.g.*, dipole moments) appear unable to account for differences among the solutes regarding their polar interactions.

The series of benzene derivatives selected as test solutes by Kaliszan *et al.* [16,17] seems to be very suitable for QSRR studies aimed at identifying of the most important factors determining RP HPLC retention. These solutes clearly differ with respect to their size and polarity. These differences were successfully quantified by E_{Ti} and Δ_i. Certainly, it would be difficult to correctly predict relative retention of two solutes of similar values of E_{Ti} and Δ_i but it is easy in case of two compounds of evidently different E_{Ti} and Δ_i.

Present considerations make it clear why the QSRR are difficult to obtain which would predict the retention (identify the peak on a chromatogram) of a member of a family of close congeners. Progress with that respect is expected with the development of molecular modeling and thus, the means of more reliable and precise structural characteristics of solutes. On the other hand, the QSRR aimed at comparison of chromatographic systems, stationary phase materials and molecular mechanisms operating during separation, can conveniently be derived provided that test series of solutes is properly designed.

The second term in square brackets in Eq. (10.27) corresponds to hypothetical $\log k$ of a solute i chromatographed on a stationary phase j in 100% methanol mobile phase, $\log k'_{Mi}$. Analysing the physical sense of the descriptors used one can conclude that $\log k'_{Mi}$ will be larger the larger is the solute as reflected by the absolute value of E_{Ti} (E_{Ti} assumes negative values; Table 10.1). E_{Ti} characterizes the ability of solute molecule to interact by dispersive interactions with molecules of both stationary and mobile phases. Net effect of these attractive interactions will provide positive contribution to retention because dispersive interactions of a solute with bulky dimethyl octadecane moieties are stronger than analogous interaction of a solute with small molecules of methanol.

Negative sign at Δ_i proves that the more polar is the solute, the lower will be its $\log k'_{Mi}$. It is because the net effect of polar interactions (dipole–dipole, dipole–induced dipole) of a solute and nonpolar hydrocarbon of the stationary phase on one hand and the polar methanol eluent on the other hand, will contribute negatively to retention. Positive sign at C_j means that $\log k'_{Mi}$ will be larger the greater is the hydrocarbon coverage of the stationary phase.

Opposite will be the effects of the descriptors E_{Ti}, Δ_i and C_j on the slope of the $\log k'_{ij}$ *vs.* X relationship (do note forget that X denotes the mole fraction of water in the binary methanol-water eluents). Thus, Eq. (10.27) is a comprehensible quantitative proof for a commonly, however intuitively, accepted view on the mechanism of RP HPLC separations.

Important observation expressed by Eq. (10.27) is that both slope and intercept of $\log k'$ *vs.* eluent composition relationship can be described by the same structural parameters. This is in agreement with the reported [19,20] relationships between such slope and intercept parameters in RP HPLC. Especially interesting appear the observations of Kaliszan *et al.* [17,18] in view of theoretical analysis of the slope-intercept relationships by Tan and Carr [20]. The latter authors concluded that some linear slope-intercept relationships are artifacts which do not reflect chemical reality. The relationships between slope and intercept which emerge from Kaliszan *et al.* [17,18] model may be more complex than simple linearity. As Tan and Carr [20] noted if linearity between slope and intercept was universal, the well-known phenomena of changes in elution sequence with the volume fraction of organic modifier in the mobile phase would not be possible. The QSRR derived by Kaliszan *et al.* [16,17] give rise to such a possibility.

10.3. Properties of Stationary Phases as Revealed by Factor Analysis of Chromatographic Data

Factorial methods of data analysis were first applied in chromatography to predict retention. Weiner and co-workers [21–24] made predictions of GC retention indices of several families of simple solutes and Huber and co-workers [25,26] first predicted liquid chromatographic partition coefficients of some steroids and pesticides. Next, the GC retention indices were the subject of factor analysis of Laffort and co-workers [27,28] Hirsch *et al.* [29] and Buydens and co-workers [30–32]. Early reports of HPLC retention prediction by a factorial approach were by Hasan and Jurs [33] and by Smith [34]. To TLC the factorial methods were introduced to classify the eluents and stationary phases [35–39].

Great impact on application of factorial methods of data analysis in chromatography had systematic studies of Chrétien and co-workers [40–43] who employed the so-called correspondence factor analysis (CFA). CFA is a variant of principal component analysis (PCA) in which the row-points (objects) and the column-points (variables) of the retention data matrix are projected simultaneously on the first factorial plane. Thus, CFA allows simultaneous solute and chromatographic system projections on the same map. The relative proximities of solutes and systems reflect the chromatographic affinity. Chrétien and co-workers [40–43] demonstrated suitability of CFA to identify best chromatographic systems for separation of a given series of solutes. They also succeeded in assigning physical meaning to some of the abstract factors extracted from retention data matrix.

In late 1980 the main interest of QSRR students, employing factorial methods to the analysis of chromatographic data, shifted from retention prediction to comparative analysis of stationary phase and to relating chromatographic behaviour of solutes to their pharmacological properties. The latter topic will be discussed in Chapter 12.

Systematic chemometric approach aimed at objective characterization of GC stationary phases was reported by Welsch *et al.* [44] and by Huber and Reich [45]. In liquid chromatography Delaney *et al.* [46] demonstrated that PCA of retention data of 10 test solutes (plus the number of theoretical plates and the degree of "free silanol groups") allowed for classification of nine octadecylsilica phases into three groups displaying similar chromatographic behaviour. PCA was also able to identify the key test compounds on which the classification of phases depended. The classification and test compound selection agreed with those accomplished by an HPLC expert.

Extensive five years long studies of the group of Prof. Jacques R. Chrétien concerning "factor analysis and experimental design in HPLC" have excellently been summarized in 1991 [47]. The PCA and CFA have been applied to a large retention data set for 63 compounds chromatographed in 43 HPLC systems. A clear-cut separation of normal-phase (NP) systems and reversed-phase (RP) systems was evident on the PCA and CFA maps. By simultaneously projecting the solutes on the planes defined by the first two principal component axes the affinity of individual compound to NP or RP systems could be determined. The multivariate approach provide rational basis for selecting a best HPLC system to separate a given pair of solutes.

Schmitz *et al.* [48] applied several multivariate methods of data analysis to a set of capacity factors, selectivities and asymmetry factors of 9 selected test solutes chromatographed on 26 RP HPLC columns. The eluent was methanol-water 55 : 45 (v/v). Seven RP-8, eleven RP-18 and one polymer-encapsulated column were the commercially available products. Seven additional polymer-coated columns were prepared by the authors [48]. Either cluster, principal component, correspondence factor and discriminant analysis showed a uniform picture: the stationary phases were always grouped in the same classes of similar properties (three classes) and the outliers were the same. Based on the multivariate analysis the stationary phases showing similar properties can be selected. For example, the polymer-coated stationary phases formed the narrowest group in CFA plots as could be expected because these columns were prepared by the same polymerization process and differed only regarding their carbon load. Schmitz *et al.* (48) were able to detect the deviation of some columns from the typical trend observed for RP-8, RP-18 or polymer-coated materials. The deviation could be explained in terms of column extreme physicochemical properties.

Kårnäs and Lindblom [43] analysed by hydrophobic interaction chromatography (HIC) a set of test proteins in a descending salt gradient using 13 gels (seven of them commercially available). The media could be divided into several groups after PCA of the test solute elution positions. It appeared that some media were selecting mainly according to the test protein hydrophobicity as expressed by the fraction of hydrophobic amino acids, and their mechanism of interaction was insensitive to protein charge. In other media, the charge on the protein was the most important factor.

Kervoort *et al.* [50] demonstrated by PCA a rational selection of the best RP HPLC columns for analysis of basic compounds. Eight commercially available columns tested included hydrocarbonaceous silica phases, alumina and zirkonia coated with polybutadiene, polyvinylalcohol-based polymer and polystyrene-divinylbenzene polymer. The multivariate analysis or retention parameters revealed that for the test basic solutes very good results were obtained at alkaline pH using a column containing zirkonium oxide particles coated with polybutadiene. At low pH values good results were obtained with new specially deactivated hydrocarbon-bound silica stationary phases (*Supelcosil LC-ABZ*, *Zorbax* $R_x - C_{18}$). Polymer-based columns showed inferior results because of the high plate heights obtained.

A recent systematic study by Hamoir *et al.* [51] concerned a set of 16 different batches, brands and types of commercially available HPLC stationary phases. Test solutes were 9 benzene derivatives. Methanol-phosphate buffer in several proportions was the eluent. To analyse retention data the authors applied a variant of PCA, the spectral mapping analysis (SMA) of Lewi [52]. In SMA the pretreatment of the data consists of a logarithmic transformation and double-centering followed by factorization and display as a biplot. Hamoir *et al.* [51] demonstrated that categorization of phases by SMA can be useful in selection of a stationary phase with similar retention properties as the one used for a specific analysis. Also, a stationary phase with completely different properties can be identified, which can be useful in method development.

Classification of a limited set of TLC and paper chromatographic systems by PCA was reported by Bazylak [53].

Typical methodology of multivariate analysis of properties of HPLC stationary phases will now be presented based on the most recent report from this author's laboratory [54]. In the work several novel silica-based, chemically bonded stationary phase materials have been synthesized and their performance tested, along with a few commercially available columns, regarding the separation of nucleosides and cyclic nucleotides. Eleven test solutes used in the study are given in Fig. 10.2.

Chemical structures of the stationary phase materials are given in Fig. 10.3. The phases OD2, 1NAPH, BN, CN, pMOB, DPH, mCF3B, 5MIND, MPH, 2PYR and 2MIM were synthesized in Prof. Bernd Jastorff's laboratory according to the previously elaborated general procedure [55,56]. The phases RP-8 (octyl silica), ALU (polybutadiene-encapsulated alumina), CARB (graphitized carbon) and IAM (immobilized artificial membrane) were commercially available materials of distinctive properties.

The capacity factors k' for 11 test solutes were determined isocratically in pH 6.0 buffer-acetonitrile mobile phase of a composition providing measurable retention data for each solute in a given HPLC system.

The test solutes were subjected to molecular modeling employing a professional software. In case of nucleosides calculations were performed for the nonionized forms

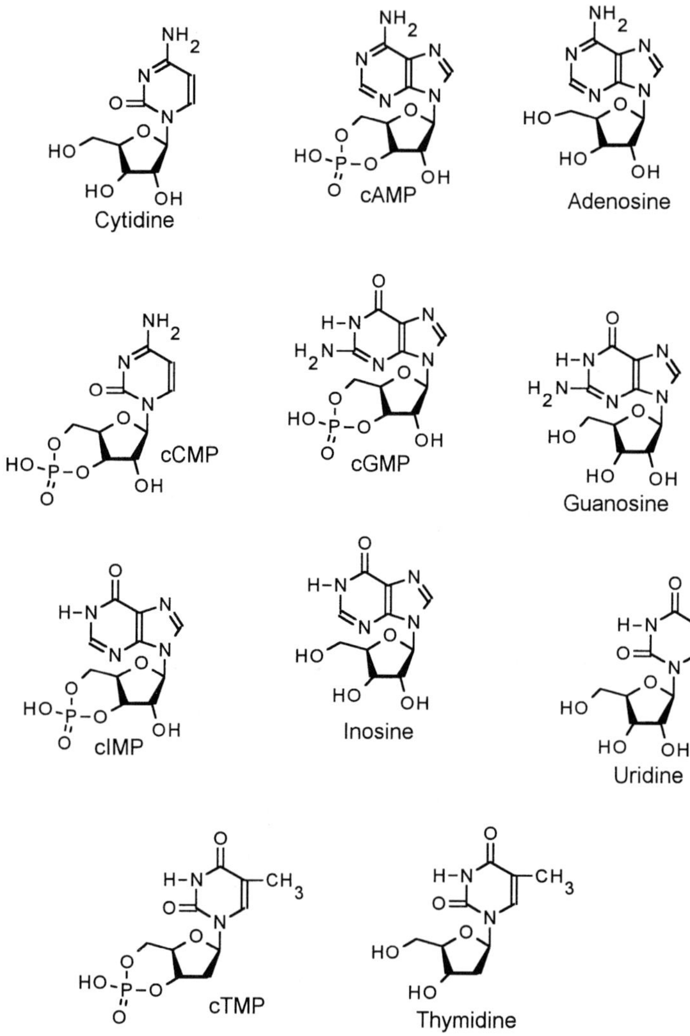

Figure 10.2. Nucleosides and cyclic nucleotides test solutes analysed by Turowski *et al.* [54].

of the solutes. In case of cyclic nucleotides respective calculations were done for the ionized (−1) forms of the solutes. For the geometry optimized structures numerous structural parameters were obtained. Selected data are collected in Table 10.2.

Selecting the structural descriptors the authors [54] presumed that molecular mass, total energy, binding energy, electronic energy and heat of formation reflect basically the differences in bulkiness among the solutes. In other words, these descriptors were expected to quantify the differences in ability of individual solutes to participate in nonspecific, dispersive intermolecular interactions with stationary phase ligands. The next two structural parameters in Table 10.2, *i.e.*, the energy of highest occupied molecular orbital (HOMO) and the energy of lowest unoccupied molecular orbital (LUMO), are considered as reflecting the abilities of solutes to take

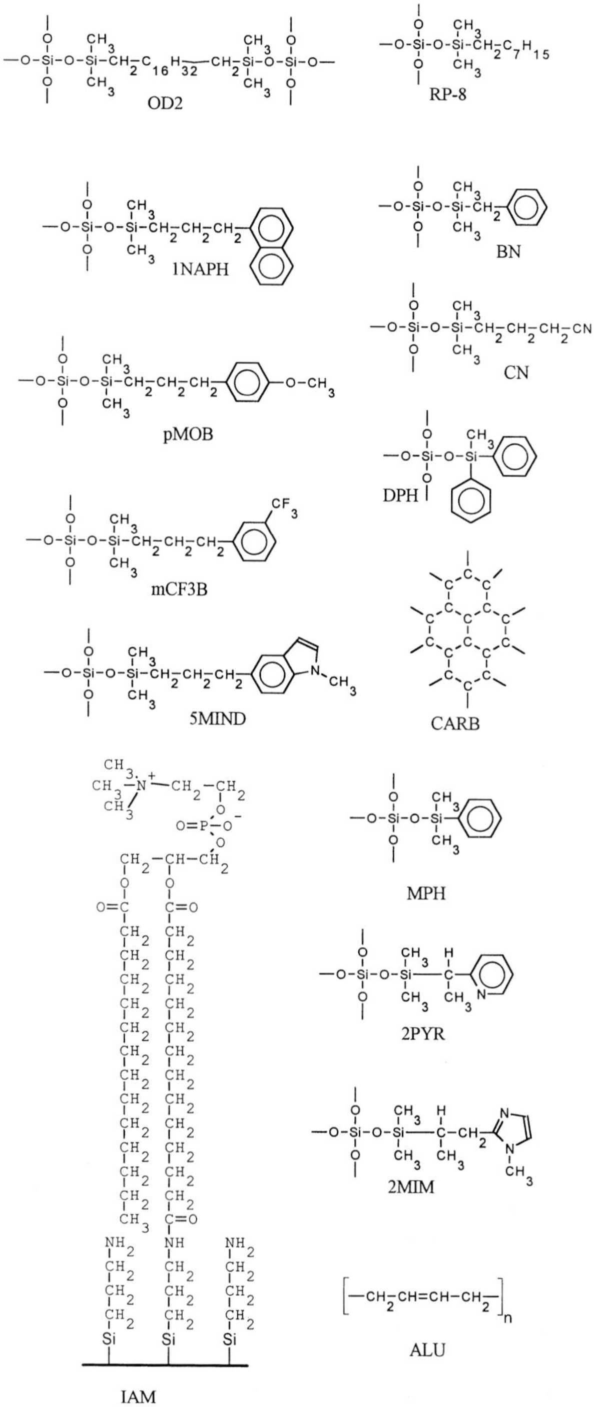

Figure 10.3. Chemical structures of stationary phases studied by Turowski *et al.* [54].

Table 10.2 Structural parameters of a series of nucleosides and cyclic nucleotides obtained from molecular modeling [54].

Solute	*Molecular Mass*	*Total Energy* [*k cal/mol*]	*Binding Energy* [*k cal/mol*]	*Electronic Energy* [*k cal/mol*]	*Heat of Formation* [*k cal/mol*]	*Energy of LUMO* [*eV*]	*Energy of HOMO* [*eV*]	*Maximum Charge* [*electrons*]	*Minimum Charge* [*electrons*]	*Dipole Moment* [*D*]
INO	268.2	−89351.6	−3217.7	−547431	−133.8	−8.9996	−0.5158	0.381	−0.335	6.61
THD	242.2	−80479.2	−3155.2	−478466	−193.1	−9.6600	−0.3404	0.408	−0.390	4.06
URD	244.2	−84280.3	−2978.4	−492188	−231.8	−10.0402	−0.4232	0.410	−0.405	3.22
CYT	243.2	−81968.7	−3025.9	−485049	−173.8	−9.3789	−0.1656	0.359	−0.366	4.21
ADO	267.2	−87056.3	−3281.5	−547501	−92.0	−8.8900	−0.2815	0.251	−0.356	1.48
GUO	283.2	−94442.4	−3378	−603868	−129.0	−8.7641	−0.4762	0.399	−0.348	5.69
cIMP (−1)	329.2	−106551	−3493.7	−673476	−319.3	−6.7569	1.5764	2.537	−1.123	13.26
cTMP (−1)	303.2	−97666.5	−3419.3	−609258	−366.7	−6.4108	2.0273	2.533	−1.138	14.91
cCMP (−1)	304.2	−99163.2	−3297.3	−607488	−354.6	−6.4881	1.8086	2.539	−1.103	18.90
cGMP (−1)	344.2	−111630	−3642.7	−726843	−303.2	−6.7370	1.5998	2.530	−1.140	13.55
cAMP (−1)	328.2	−104241	−3543.3	−670199	263.4	−6.6709	1.7611	2.532	−1.141	16.14

part in the electron pair donor-electron pair acceptor interactions and/or hydrogen bonding interactions. Maximum excess charge, minimum excess charge on individual atoms in the solute molecule and dipole moment are presumed to quantify differences among the solutes regarding the electrostatic interactions of the type dipole–dipole and dipole–induced dipole.

The 11×15 matrix of capacity factors determined for 11 solutes in 15 HPLC systems was subjected to PCA employing a commercially available professional statistical software. Similarly, PCA was done on the 11×10 matrix of structural parameters of the solutes. In case of retention data, the first principal component (PC1) accounted for 79.6% and the second principal component (PC2) for 13.7% of data variance. The inputs to PC1 and PC2 due to the individual HPLC systems and due to the solutes are displayed in Figs. 10.4 ("loadings") and 10.5 ("scores").

In case of structural descriptors, the first principal component (PC1) accounted for 86.6% and the second principal component (PC2) for 10.4% of total data variance.

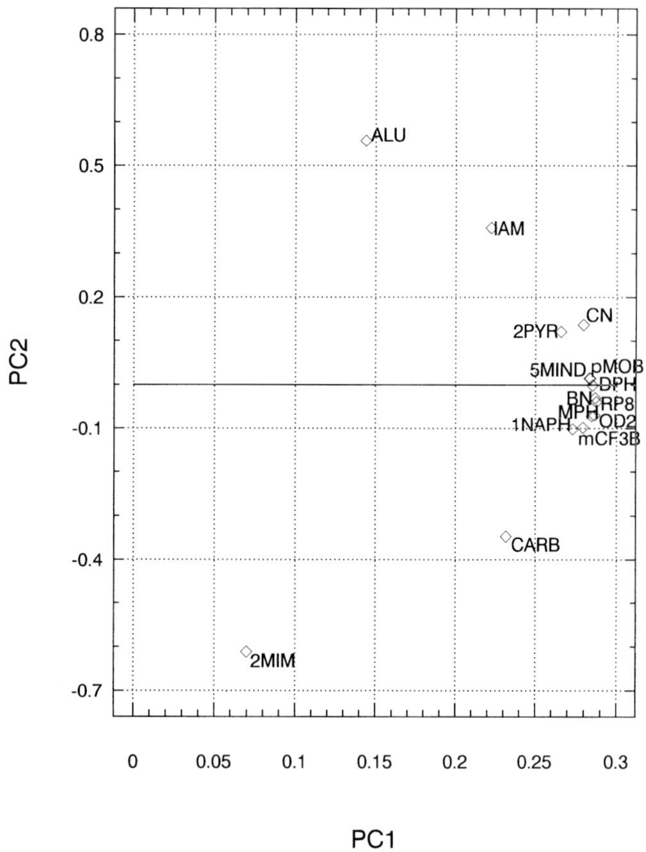

Figure 10.4. Plot of inputs ("loadings") to first two principal components (PC1 and PC2) obtained from the analysis of capacity factors of 11 nucleosides and cyclic nucleotides on 15 stationary phases due to individual stationary phases [54].

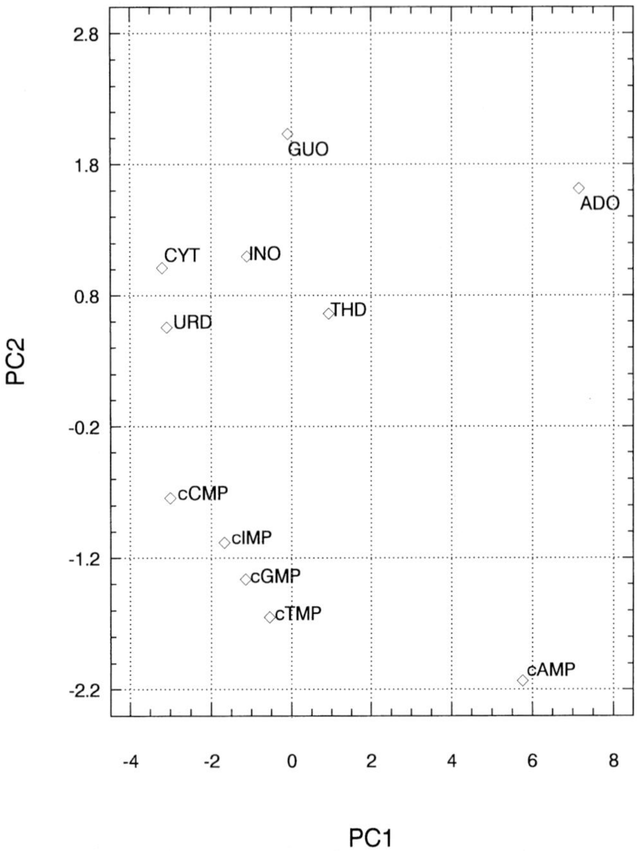

Figure 10.5. Plot of inputs ("scores") to first two principal components (PC1 and PC2) obtained from the analysis of capacity factors of 11 nucleosides and cyclic nucleotides on 15 stationary phases due to individual solutes [54].

The inputs to PC1 and PC2 by the descriptors and by individual solutes are presented in Figs. 10.6 ("loadings") and 10.7 ("scores"), respectively.

Distribution of 15 HPLC systems in Fig. 10.4 reveals a compact cluster of hydrocarbonaceous silica-based stationary phases. Close distances between points corresponding to 1NAPH, mCF3B, OD2, MPH, BN, DPH and RP-8 indicate similar discriminative properties of the phases toward the test nucleosides and cyclic nucleotides. There is no much difference in the separation properties between the aromatic and aliphatic hydrocarbon phases. To the same cluster belong also the phases pMOB and 5MIND. Evidently, single heteroatoms (ether oxygen in pMCB and tertiary nitrogen in 5MIND) do not change the overwhelmingly hydrocarbonaceous character of these phases. On the other hand, the cyano functionality in CN and pyridine moiety in 2PYR provide specific interactions which are significant enough to separate these phases from the cluster of purely hydrocarbonaceous phases. The most striking specificity can be assigned to the imidazoline derivative phase 2MIM. It appears that 2MIM has an enhanced affinity to the ionised forms

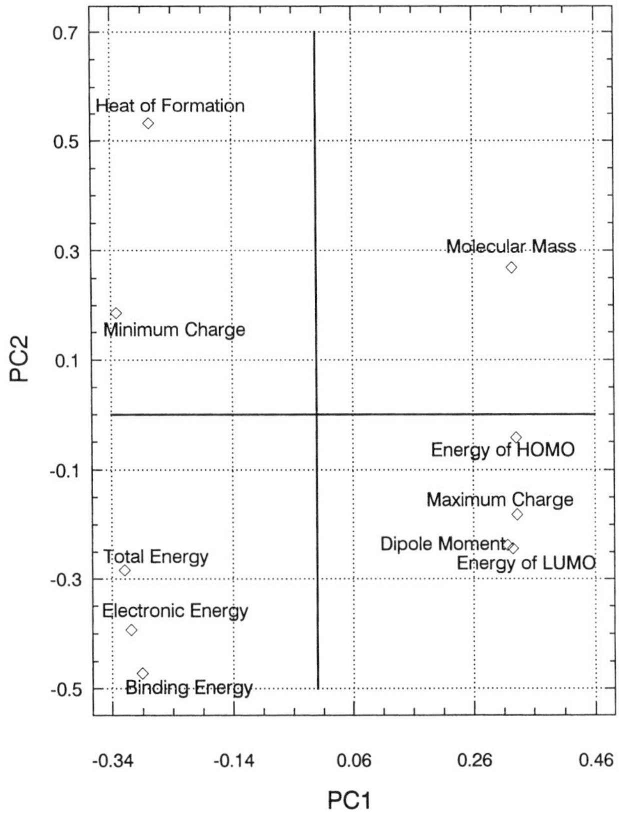

Figure 10.6. Plot of inputs ("loadings") to first two principal components (PC1 and PC2) obtained from the analysis of structural descriptors of solutes (Table 10.2) due to individual structural parameters [54].

of cyclic nucleotides. Opposite properties have the phases ALU and IAM: they seem to preferentially retain nucleosides.

Conclusions drawn from distribution of stationary phases (variables) are supported by the distribution of test solutes (objects) in Fig. 10.5. It must be noted here that the assigning of physical meaning to complex abstract factors is in most instances not straightforward. It appears that positive value of PC2 is typical for basic nucleosides whereas negative values of PC2 are observed for acidic cyclic nucleotides. Thus, PC2 can be assumed to extract information on acid-base properties of solutes. Regarding PC1, the highest inputs to it provide ADO and cAMP and the lowest CYT, URD and cCMP. Referring to a standard measure of hydrophobicity, *i.e.*, logarithm of *n*-octanol-water partition coefficient ($\log P$), available for two nucleosides [57], one can assume that PC1 extracts information on hydrophobicity of solutes: $\log P$ for ADO is -1.10 and for URD $\log P$ is -1.98.

As far as the newly synthesized stationary phases are concerned the 2MIM material appears especially interesting from the point of view of separation of nucleosides and cyclic nucleosides. Also, 2PYR and CN may provide distinguishable

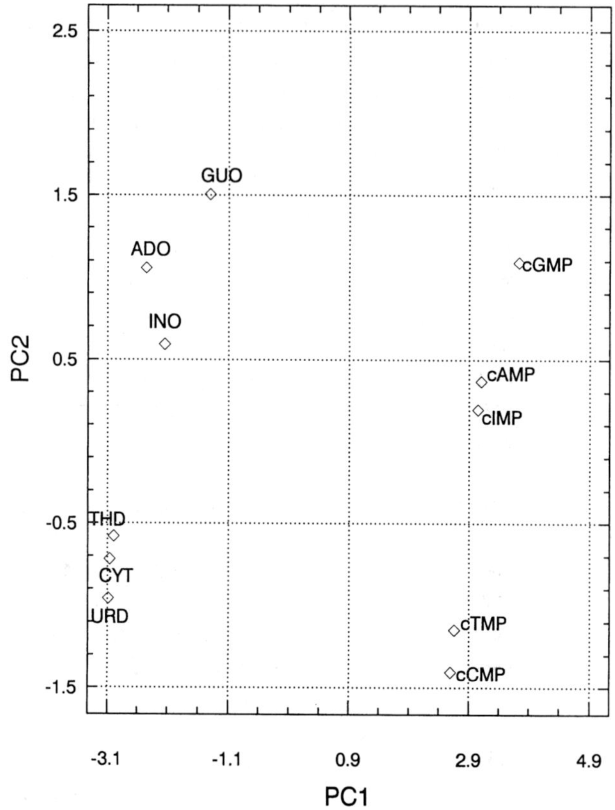

Figure 10.7. Plot of inputs ("scores") to first two principal components (PC1 and PC2) obtained from the analysis of structural descriptors of solutes (Table 10.2) due to individual solutes [54].

separation patterns. Other new phases seem to have no clear-cut advantages over the classical hydrocarbonaceous materials.

To further elucidate the mechanism of HPLC separations of nucleosides and cyclic nucleosides a set of molecular descriptors of solutes was subjected to PCA. As illustrated in Fig. 10.6 PC1 is determined mostly by electronic descriptors: maximum and minimum atomic excess charges, energies of HOMO and LUMO, and dipole moment. PC2 is determined by structural descriptors related to molecular size: heat of formation, molecular mass, and the total, electronic and binding energies.

Conclusions drawn from Fig. 10.6 are confirmed by the PCA plot for the solutes (Fig. 10.7). As evident from Fig. 10.7, polar cyclic nucleotides have high values of PC1 and nucleosides possess low values of PC1. Distribution of nucleosides and cyclic nucleotides along the PC2 axis is according to their molecular mass (provided that two subgroups are considered separately). However, PC2 is not a simple function of molecular mass because heavier cyclic nucleotides have lower values of PC2 than their corresponding nucleosides.

Turowski *et al.* [54] summarize their study as follows; (i) of 11 newly synthesized stationary phase materials 8 hydrocarbon-bonded silica phases have separating properties closely similar to a standard, commercially available octylsilica phase. The phases comprising aromatic fragments in their ligands do not differ significantly from the aliphatic hydrocarbon materials; (ii) New stationary phases comprising the methylimidazole, pyridine and cyano moiety in their structure have specific, distinguishable properties regarding the separation of nucleosides and cyclic nucleotides; (iii) PCA of retention parameters provides an objective quantitative means to mutually compare separation properties of a large number of diversified HPLC systems; (iv) PCA of structural descriptors of solutes obtained by molecular modeling facilitates the identification of structural features that most strongly affect retention in individual HPLC systems.

10.4. Molecular Mechanism of Chromatographic Retention in view of Multivariate Analysis of Chromatographic and Structural Data

Soon after Knox *et al.* [58] introduced porous graphitic carbon as an HPLC stationary phase, it appeared that the retention behaviour observed was different to that expected for a typical reversed-phase material. The carbon phase can also be operated at normal-phase conditions with heptane as an eluent. For that operation mode, a classical QSRR study based on multiple regression analysis [59] demonstrated that specific, polar, molecular size-independent solute properties were decisive for retention. To further study the mechanism of HPLC retention on graphitic carbon Kaliszan *et al.* [60] applied the PCA method to a set of 19 structural descriptors of 20 simple aromatic test solutes.

The total of 19 descriptors included molecular weight (molwt), refractivity (bondrefr), Kier and Hall [61] generalized molecular connectivity index of second order (chi2v), kappa indices of Kier [62] of second and third order (kappa2, kappa3), the average distance sum connectivity index of Balaban (balaban) calculated according to Barysz *et al.* [63], the information content indices of neighbourhood symmetry of zeroth, first and second order (ic0, ic1 and ic2, respectively) [64]. The separate group of nonempirical structural descriptors were quantum chemical indices calculated for each solute by using the MNDO3 method [55] on molecular coordinate files obtained form the Cambridge Crystallographic Database. This group of descriptors included total energy (etotal), heat of formation (heatf), energy of core–core repulsion (corer), energy of highest occupied molecular orbital (ehomo), energy of lowest unoccupied molecular orbital (elumo) and dipole moment (dipolem). Also electron excess charges on individual atoms were calculated. From these, the first order submolecular polarity parameter was calculated as the largest difference of excess charges for a pair of atoms in a molecule (delta1) [66]. The second-order submolecular polarity parameter, delta2, was determined analogously as the second largest difference of excess charges. The parameter diploc after Kikuchi [67] and the geometric-electronic index (geoeli) after Ośmiałowski *et al.* [68] were also calculated based on electron excess charge distribution within a solute molecule.

Logarithms of HPLC capacity factors, $\log k'$, of test solutes are given in Table 10.3.

Principal component analysis (PCA) of the 20×19 solute descriptors matrix produced the first principal component, PC1, accounting for 57% of total variance within original structural data set, the second principal component, PC2, accounting

Table 10.3 Retention data, log *k*′, determined on porous carbon with heptane eluent, for a series of aromatic test solutes [60].

No.	*Solvies*	*log k′*
1	Phenol	0.431
2	*p*-Cresol	0.456
3	4-Ethyl phenol	0.277
4	4-*n*-Propyl phenol	0.312
5	2-Isopropyl phenol	−0.171
6	2-tert-Butyl phenol	−0.419
7	4-tert-Butyl phenol	−0.022
8	Carvacrol	−0.194
9	Thymol	−0.148
10	Benzene	−1.060
11	Toluene	−0.791
12	*o*-Xylene	−0.433
13	*m*-Xylene	−0.666
14	*p*-Xylene	−0.658
15	Anisole	−0.317
16	Benzyl alcohol	−0.211
17	Chlorobenzene	−0.606
18	Acetophenone	0.283
19	Nitrobenzene	0.799
20	Methyl benzoate	0.236

for 21% variance and the third principal component accounting for 7% of total variance. The "loadings" by individual structural parameters were subjected to a VARIMAX rotation. The inputs by the descriptors to the rotated principal components, PCvr1 and PCvr2, are graphically depicted (Fig. 10.8).

As can be noticed in Fig. 10.8 the first principal component, PCvr1, is "loaded" mostly by such structural descriptors as refractivity (bondrefr), energy of core–core repulsion (corer), valence connectivity index of second order (chi2v), Balaban index (balaban), molecular weight (molwt), kappa index of third order (kappa3) and total energy (etotal). To a lesser extent is PCvr1 "loaded" by kappa2, geometric-electronic index (geoeli) and information content indices of first and second order, ic1 and ic2. It is evident that all these structural descriptors reflect basically the bulkiness of the solutes. In such a situation, PCvr1 condenses information on molecular size.

The second principal component, PCvr2, is "loaded" predominantly by the following structural descriptors: dipole moment (dipolem), energy of LUMO (elumo) information content index of seroth order (ic0) and submolecular polarity parameters of first and second order (delta1 and delta2). Hence it may be concluded that PCvr2 concentrates structural information related to molecular polarity. Specific, polar properties of chemical compounds are the result of electron distribution within a molecule. From the chromatographic point of view, such properties determine the ability of a solute to participate in intermolecular interactions with the stationary and mobile phase of the dipole–dipole, dipole–induced dipole and electron pair donor-electron pair acceptor type. Some aspects of solute polarity are comprised in PCvr3 which is "loaded" by local dipole index (diploc) and energy of HOMO (ehomo).

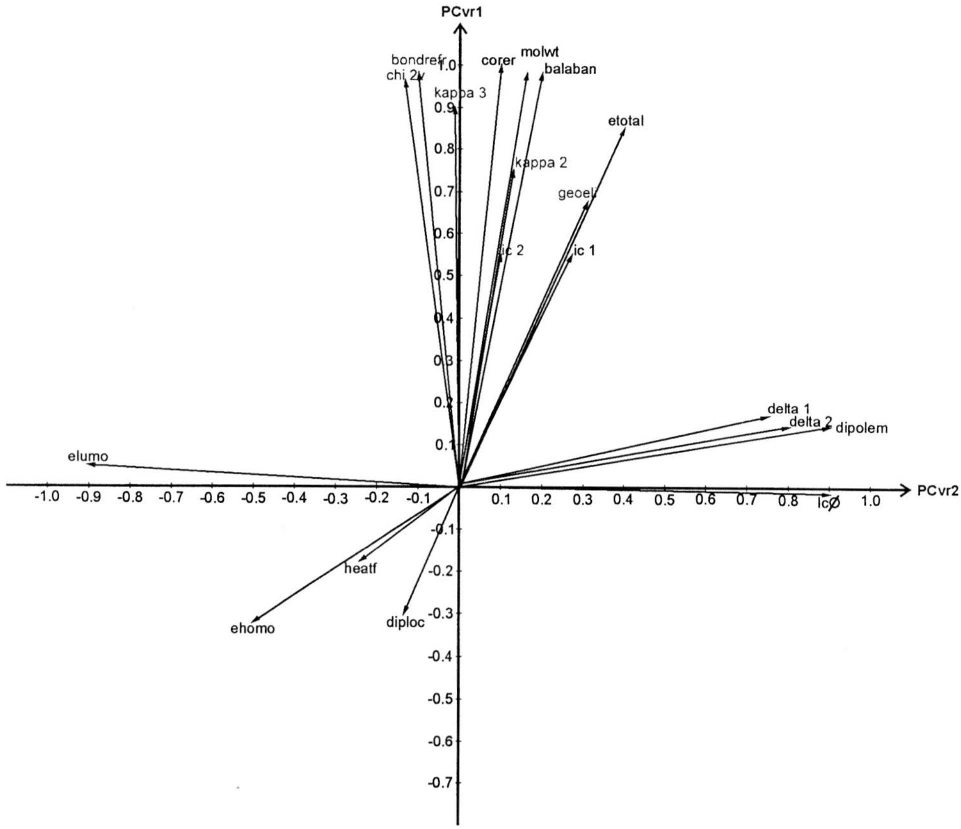

Figure 10.8. VARIMAX-rotated "loadings" of two first principal components by individual structural descriptors considered by Kaliszan *et al.* [60]. See text for the explanation of the symbols. (Reprinted from Kaliszan, R., Ośmiałowski, K., Bassler, B. and Hartwick, R.A., *J. Chromatogr.* 499, 333, 1990 with kind permission of Elsevier Science-NL, Sara Burgerhartstraat 25, 1055 KV Amsterdam, The Netherlands.)

The principal component scores (objects) after VARIMAX rotation differentiated a subgroup of phenolic solutes from the remaining benzene derivatives.

The principal component regression analysis yielded the following equation describing $\log k'$ of the whole set of solutes:

$$\log k' = -0.145 - 0.323 \text{ PCvr2} + 0.259 \text{ PCvr3}; \qquad (10.28)$$

$$n = 20; \quad R = 0.85; \quad \text{s} = 0.271; \quad F = 22.$$

On analysing Eq. (10.28), one notes the absence of the molecular size-related PCvr1. As well known, the bulkiness of chemical compounds determines their ability to participate in nonspecific dispersive interactions. The lack of significance of PCvr1 for a description of the retention of the whole set of diverse solutes can be interpreted that the net London-type attraction of solutes and stationary phase is comparable to that between solutes and mobile phase molecules. It can also be that the input by dispersive attraction to $\log k'$ does not differ statistically within the set of solutes studied. Anyway, the bulkiness of solutes is not decisive for their retention on carbon with heptane as the eluent.

The PCvr2 term in Eq. (10.28) can be interpreted as comprising the charge-transfer attractive interactions of the solutes and graphitic carbon on the one hand, and the solute-heptane inductive interactions on the other. The contribution of PCvr3 to Eq. (10.28) is difficult to interpret in terms of the retention mechanism. Bearing in mind the input by the local dipole index (diploc) to PCvr3 one can consider it as reflecting dipole-induced dipole interactions involving solute molecules and the molecules of both chromatographic phases.

Equation (10.28) gives some insight into mechanism of separation in the graphitic carbon-heptane HPLC systems. Its statistical value, however, is low for precise predictions of retention. In the original paper [60] one can find highly predictive QSRR for subgroups of solutes. Equation (10.28) is important because it summarizes main features of retention in rather unique HPLC separation system.

The findings of Kaliszan *et al.* [60], summarized by Eq. (10.28) for porous graphitic carbon-heptane system, were confirmed by Forgacs and co-workers [69,70]. These authors concluded from their PCA of retention data of phenol and aniline derivatives, determined in reversed-phase mode (methanol-water and acetonitrile-water eluents), that the retention was not governed by solute bulkiness and lipophilicity. The lack of correlation of retention data determined on porous carbon with lipophilicity is in agreement with the earlier report by De Biasi *et al.* [71].

Specific retentive properties of graphitic carbon were observed by Tanaka *et al.* [72]. Excellent proof for a typical behaviour of porous graphitic carbon in reversed-phase systems with water-rich mobile phases was recently provided by Hennion *et al.* [73]. These authors obtained correlations of their $\log k'$ extrapolated to aqueous mobile phase ($\log k'_w$) with octanol-water $\log P$ for solutes containing hydrophobic substituents (alkyl or chloro groups). However, for aromatic solutes containing polar substituents there were no such correlations. Instead, the retention factor increased with increase in the number of polar substituents and was shown to depend on both the field and the mutual resonance effects of the different substituents on the aromatic ring. Hennion *et al.* [73] conclude that electronic interactions are more important than hydrophobic interactions in the retention mechanism of polar compounds. These authors parametrized the polarity of the solutes using local dipolar moments as well as the overall electron-excess charge density and were able to predict retention.

The lack of significance of solute bulkiness for normal-phase retention on graphitized carbon was again confirmed by a QSRR analysis of $\log k'$ parameters of a series of benzodiazepine drugs [74].

Properties of another new stationary phase material, *i.e.*, polybutadiene-coated alumina (PBCA) appear quite different from graphitized carbon in view of the multivariate analysis of retention and structural data [75]. The PBCA stationary phase, introduced by Schomburg and co-workers [76], can be operated over a wide range of pH, including alkaline environments. Hence, in contrast to the standard silica-based reversed-phase materials, basic solutes can be chromatographed in nonionized form. Another feature of the classical RP HPLC materials is a presence of free silanol sites which obviously is not the case with PBCA.

For a specially designed set of 21 solutes (see Section 9.1, Fig. 9.1) retention data, extrapolated to pure buffer, $\log k'_w$, for non-ionized forms of acids, bases and neutrals were determined on PBCA with methanol-buffer eluents. Structural descriptors were generally the same types as the above discussed descriptors of solutes chromatographed on graphitized carbon. The PCA of a 21×16 structural data matrix yielded the first principal component accounting for 48.6% of total data variance and the second principal component accounting for 25.2% variance. The principal components were subjected to VARIMAX rotation. The PCvr1 was found to be "loaded" mostly by the first-order and second-order valence connectivity indices, Wiener index, molar refractivity, molecular mass and total energy. All these structural descriptors reflect the bulkiness of solutes. This means that PCvr1 condenses information on solute size.

The second VARIMAX-rotated principal component was "loaded" predominantly by dipole moment, the submolecular polarity parameter Δ and information content indices of zeroth-, first- and second-order. Hence it can be concluded that PCvr2 concentrates structural information related to molecular polarity.

Principal component (non-rotated) regression analysis resulted in the following equation:

$$\log k'_w = 0.8852 + 0.5942\,\mathrm{PC1} - 0.9016\,\mathrm{PC2}; \tag{10.29}$$

$$n = 21; \quad R = 0.948; \quad s = 0.380; \quad F = 81.$$

Both PC1 and PC2 terms in Eq. (10.29) are significant at least the 0.001 significance level. The relationship between experimental $\log k'_w$ data and those calculated by Eq. (10.29) is illustrated in Fig. 10.9.

The interpretation of Eq. (10.29) is straight forward: the increasing solute bulkiness enhances retention whereas increasing polarity has opposite effect. Evidently, non-specific, dispersive interactions of solutes with the stationary phase hydrocarbons prevail over the analogous interactions with the mobile phase. This is indicated by the positive increment to Eq. (10.29) by structural parameters reflecting the ability of solutes to participate in London-type intermolecular interactions. On the other hand, the negative increment to Eq. (10.29) of the molecular parameters reflecting the ability of solutes to participate in specific, polar intermolecular interactions demonstrates that polar solute-mobile phase interactions are stronger than analogous solute-stationary phase interactions. Hence the mechanism of retention on PBCA is closely similar to that on octadecylsilica (ODS) [16,17] but basically different from that postulated for HPLC on porous graphitic carbon.

The PCA was also applied in studies of retention mechanism operating in HPLC (and TLC) on alumina columns (layers) using *n*-hexane-dioxane eluents [77] demonstrating similarity (expected) of the systems. Similar, however "slightly different", appeared also retention behaviour of polyethylene-coated silica and octadecylsilica HPLC column with regards to aniline derivative solutes [78].

By means of PCA, Chrétien and co-workers [79] demonstrated a single retention mechanism operating in ion chromatography of test carboxylate anions with 25 carbonate eluents. This mechanism is the ion-exchange equilibrium.

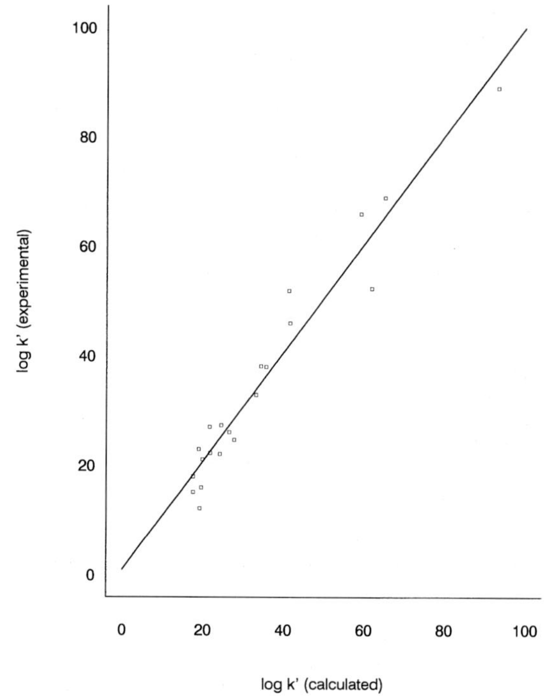

Figure 10.9. Relationship between log k' (extrapolated to pure buffer) determined experimentally and calculated by Eq. (10.29). (Reprinted from Kaliszan, R. and Ośmiałowski, K. *J. Chromatogr.* **506**, 3, 1990 with kind permission of Elsevier-Science-NL, Sara Burgerhartstraat 25, 1055 KV Amsterdam, The Netherlands).

A comparative PCA study of HPLC and supercritical fluid chromatography (SFC) retention data determined on various bonded silica columns demonstrated that neither of the two systems clearly differed in the retention mechanism and the separation abilities [80].

References

1. Kaliszan, R., in: "*High Performance Liquid Chromatography*", Brown, P.R. and Hartwick, R.A. (Eds.), Wiley, New York, 1989, Chapter 14, p. 563.
2. Kaliszan, R. and Höltje, H.-D., *J. Chromatogr.* **234**, 303 (1982).
3. Grzybowski, J., Lamparczyk, H., Nasal, A. and Radecki, A., *J. Chromatogr.* **196**, 217 (1980).
4. Hansch, C., Leo, A., Unger, S.H., Kim, K.-H., Nikaitani, D. and Lien, E.J., *J. Med. Chem.* **16**, 1207 (1973).
5. Kaliszan, R., *J. Chromatogr.* **220**, 71 (1981).
6. Halkiewicz, J., Lamparczyk, H. and Radecki, A., *Fresenius Z. Anal. Chem.* **320**, 577 (1985).
7. Kaliszan, R., "*Quantitative Structure-Chromatographic Retention Relationships*", Wiley, New York, 1987, p. 182.
8. Bermejo, J. and Guillén, M.D., *J. Chromatogr.* **318**, 187 (1985).
9. Bermejo, J., Blanco, C.G. and Guillén, M.D., *J. Chromatogr.* **331**, 237 (1985).
10. Bermejo J., Blanco, C.G. and Guillén, M.D., *J. Chromatogr.* **351**, 425 (1986).
11. Kleinert, T. and Ecknig, W., *J. Chromatogr.* **315**, 75 (1984).

12. Kleinert, T., Ecknig, W. and Novàk, J., *J. Chromatogr.* **315**, 851 (1984).
13. Jinno, K. and Kawasaki, K., *Chromatographia* **18**, 90 (1984).
14. Jinno, K. and Kawasaki, K., *J. Chromatogr.* **298**, 326 (1984).
15. Jinno, K. and Kawasaki, K., *J. Chromatogr.* **316**, 1 (1984).
16. Kaliszan, R., Ośmiałowski, K., Tomellini, S.A., Hsu, S.-H., Fazio, S.D. and Hartwick, R.A., *Chromatographia* **20**, 705 (1985).
17. Kaliszan, R., Ośmiałowski, K., Tomellini, S.A., Hsu, S.-H., Fazio, S.D. and Hartwick, R.A., *J. Chromatogr.* **352**, 141 (1986).
18. Kimata, K., Hosoya, K., Araki, T., Tanaka, N., Barnhart, E.R., Alexander, L.R., Sirimanne, S., McClure, P.C., Grainger, J. and Patterson, D.G. Jr., *Anal. Chem.* **65**, 2502 (1993).
19. Valkó, K., Snyder, L.R. and Glajch, J.L., *J. Chromatogr. A* **656**, 501 (1993).
20. Tan, L.C. and Carr, P.W., *J. Chromatogr. A* **656**, 521 (1993).
21. Weiner, P.H., Dack, C.J. and Howery, D.G., *J. Chromatogr.* **69**, 249 (1972).
22. Weiner, P.H. and Howery, D.G., *Can. J. Chem.* **50**, 448 (1972).
23. Weiner, P.H. and Parcher, J.F., *J. Chromatogr. Sci.* **10**, 612 (1972).
24. Kindsvater, J.H., Weiner, P.H. and Klingen, T.J., *Anal. Chem.* **46**, 982 (1974).
25. Huber, J.F.K., Alderlieste, E.T., Harren, H. and Poppe, H., *Anal. Chem.* **45**, 1337 (1973).
26. Huber, J.F.K., Meijers, C.A.M. and Hulsman, J.A.R., *J. Anal. Chem.* **44**, 111 (1982).
27. Laffort, P. and Patte, F., *J. Chromatogr.* **126**, 625 (1976).
28. Patte, F., Etcheto, M. and Laffort, P., *Anal. Chem.* **54**, 2239 (1982).
29. Hirsch, R.F., Gaydosh, R.J. and Chrétien, J.R., *Anal. Chem.* **52**, 723 (1980).
30. Buydens, L. and Massart, D.L., *Anal. Chem.* **53**, 1990 (1981).
31. Buydens, L., Massart, D.L. and Geerlings, P., *Anal. Chem.* **55**, 738 (1983).
32. Buydens, L., Coomans, D., Vanbelle, M., Massart, D.L. and Vanden Driessche, R., *J. Pharm. Sci.* **72**, 1327 (1983).
33. Hasan, M.N. and Jurs, P.C., *Anal. Chem.* **55**, 263 (1983).
34. Smith, R.M., *Anal. Chem.* **56**, 256 (1984).
35. Cserháti, T. and Bordas, B., *J. Chromatogr.* **286**, 131 (1984).
36. Cserháti, T., Bordas, B., Ekiert, L. and Bojarski, J., *J. Chromatogr.* **287**, 385 (1984).
37. Musumarra, G., Scarlata, G., Romano, G., Clementi, S. and Wold, S., *J. Chromatogr. Sci.* **22**, 538 (1984).
38. Musumarra, G., Scarlata, G., Cirma, G., Romano, G., Plazz, S., Sergio, C. and Giulietti, G., *J. Chromatogr.* **295**, 31 (1984).
39. Valko, K., Olajos, S. and Cserháti, T., *J. Chromatogr.* **499**, 361 (1990).
40. Chrétien, J.R., Szymoniak, K., Lion, C. and Haken, J.K., *J. Chromatogr.* **324**, 355 (1985).
41. Walczak, B., Dreux, M., Chrétien, J.R., Szymoniak, K., Lafosse, M., Morin-Allory, L. and Doucet, J.P., *J. Chromatogr.* **353**, 109 (1986).
42. Walczak, B., Chrétien, J.R., Dreux, M., Morin-Allory, L., Lafosse, M., Szymoniak, K. and Membrey, F., *J. Chromatogr.* **353**, 123 (1986).
43. Szymoniak, K. and Chrétien, J.R., *J. Chromatogr.* **404**, 11 (1987).
44. Welsch, T., Frank, H.M., Zwanziger, H., Liebisch, S. and Engewald, W., *Chromatographia* **19**, 457 (1984).
45. Huber, J.F.K. and Reich, G., *J. Chromatogr.* **294**, 15 (1984).
46. Delaney, M.F., Papas, A.N. and Walters, M.J., *J. Chromatogr.* **410**, 31 (1987).
47. Righezza, M. and Chrétien, J.R., *J. Chromatogr.* **556**, 169 (1991).
48. Schmitz, S.J., Zwanziger, H. and Engelhardt, H., *J. Chromatogr.* **544**, 381 (1991).
49. Kårsnäs, P. and Lindblom, T., *J. Chromatogr.* **599**, 131 (1992).
50. Kervoort, R.J.M., Derksen, M.W.J. and Maris, F.A., *J. Chromatogr. A* **678**, 1 (1994).
51. Hamoir, T., Cuesta Sánchez, F., Bourguignon, B. and Massart, D.L., *J. Chromatogr. Sci.* **32**, 488 (1994).
52. Tielemans, A., Lewi, P.J. and Massart, D.L., *Chemom. Intell. Lab. Syst.* **3**, 277 (1988).
53. Bazylak, G., *J. Planar Chromatogr.* **7**, 428 (1994).
54. Turowski, M., Kaliszan, R., Lüllmann, C., Genieser, H.G. and Jastorff, B., *J. Chromatogr. A* **728**, 201 (1996).
55. Genieser, H.G., Gabel, D. and Jastorff, B., *J. Chromatogr.* **269**, 127 (1983).
56. Lüllmann, C., Genieser, H.G. and Jastorff, B., *J. Chromatogr.* **354**, 434 (1986).
57. Craig, P.N., in: "*Comprehensive Medicinal Chemistry*", Vol. 6, Hansch, C., Sammes, P.G. and Taylor, J.B. (Eds.), Pergamon Press, Oxford, 1990.
58. Knox, J.H., Kaur, B. and Milward, G.R., *J. Chromatogr.* **352**, 3 (1986).
59. Bassler, B., Kaliszan, R. and Hartwick, R.A., *J. Chromatogr.* **461**, 139 (1989).
60. Kaliszan, R., Ośmiałowski, K., Bassler, B. and Hartwick, R.A., *J. Chromatogr.* **499**, 333 (1990).
61. Kier, L.B. and Hall, L.H., *J. Pharm. Sci.* **65**, 1806 (1976).

62. Kier, L.B., *Acta Pharm. Jugosl.* **36**, 171 (1986).
63. Barysz, M., Jashari, G., Lall, R.S., Srivastava, V.K. and Trinajstić, N., in: "*Chemical Applications of Topology and Graph Theory*", King, R.B. (Ed.), Elsevier, Amsterdam, 1983, p. 222.
64. Sarkar, R., Roy, A.B. and Sarkar, P.K., *Math. Biosci.* **39**, 299 (1978).
65. Dewar, M.J.S., *QCPE* 11, 279 (1975).
66. Ośmiałowski, K., Halkiewicz, J., Radecki, A. and Kaliszan, R., *J. Chromatogr.* **346**, 53 (1985).
67. Kikuchi, O., *Quant. Struct.-Act. Relat.* **6**, 179 (1987).
68. Ośmiałowski, K., Halkiewicz, J. and Kaliszan, R., *J. Chromatogr.* **361**, 63 (1986).
69. Forgács, E., Cserháti, T. and Valkó, K., *J. Chromatogr.* **592**, 75 (1992).
70. Forgács E. and Cserháti, T., *J. Chromatogr.* **600**, 43 (1992).
71. De Biasi, V., Lough, W.J. and Evans, M.B., *J. Chromatogr.* **353**, 279 (1986).
72. Tanaka, N., Tanigawa, T., Kimata, K., Hosoya, K. and Araki, T., *J. Chromatogr.* **549**, 29 (1991).
73. Hennion, M.-C., Coquart, V., Guenu, S. and Sella, C., *J. Chromatogr. A* **712**, 287 (1995).
74. Kaliszan, R., Kaliszan, A., Noctor, T.A.G., Purcell, W.P. and Wainer, I.W., *J. Chromatogr.* **609**, 69 (1992).
75. Kaliszan, R. and Ośmiałowski, K., *J. Chromatogr.* **506**, 3 (1990).
76. Bien-Vogelsang, K., Deege, A., Figge, H., Kohler, J. and Schomburg, G., *Chromatographia* **19**, 170 (1984).
77. Cserháti, T. and Forgács, E., *J. Chromatogr. A* **668**, 495 (1994).
78. Forgács, E., *J. Liq. Chromatogr.* **16**, 3757 (1993).
79. Haldna, U., Pentchuk, J., Righezza, M. and Chrétien, J.R., *J. Chromatogr. A* **670**, 51 (1994).
80. Morin-Allory, L. and Herbreteau, B., *J. Chromatogr.* **590**, 203 (1992).

11 Quantitative Relationships between Hydrophobicity of Solutes and their Chromatographic Retention

11.1. Historical Overview

Classical works of Overton [1], Meyer [2] and Baum [3], published a century ago, established the view that the lipophilic properties of xenobiotics are of importance in their pharmacological (toxicological) activity. In 1950 Martin [4] linearly related the partition chromatography parameter, R_M, and the logarithm of partition coefficient, $\log P_s$, determined in a system similar to the chromatographic system. Some 15 years latter first papers appeared in which hydrophobicity of drugs has been assessed by TLC [5–7]. Since early 1970s reversed-phase HPLC became the dominating method of hydrophobicity determination [8]. By now, hundreds of papers have been published dealing with chromatographic parametrization of hydrophobic properties of solutes and application of the parameters obtained in correlation studies with biological activity. The literature has been reviewed systematically since 1975 [9]. Earlier original papers and review articles have been discussed and referenced in 1987 [10]. Since that time additional reviews have been published in internationally distributed journals [11–21]. Readers are encouraged to confront this author's treatment of the problem with other authors' views [11–13,20,21].

11.2. Hydrophobicity Concept

There has been much discussion on the meaning of the term "hydrophobicity". Hydrophobicity is as good a term as any with which to describe particular temperature dependence of the solvation of nonpolar solutes in water [22]. The particular feature is that nonpolar solvation in water is strongly opposed by entropy at around 25°C and has a large positive heat capacity change. Simpler solvation processes are instead opposed by enthalpy over a wide range of temperature and have small heat capacity changes.

Commonly, hydrophobicity is understood as a measure of the relative tendency of a solute to prefer a nonaqueous over an aqueous environment or as a measure of the tendency of two (or more) solute molecules to aggregate in aqueous solutions [23]. Both measures express some kind of "phobia" for the aqueous environment. However complex, the hydrophobic properties are the net effect of the same well known physicochemical interactions that determine the state of all matter, *i.e.*, orientation, inductive, dispersive, hydrogen bonding and charge transfer interactions [24]. The problem with hydrophobic interactions is that chemists have been accustomed to the supposition that a force between two particles is a property of the particles *themselves*. In case of hydrophobic interactions the forces appear to depend mainly on the properties of the solvent and not on the solutes.

Whatever the hydrophobic effect really is, it is assumed to be the "driving force" for liquid–liquid partitioning, for passive transport through biological membranes and for drug-receptor binding. Thus, pharmacokinetic processes of drug absorption into systemic circulation, distribution among various body compartments and excretion, all involving penetrations through lipid membranes and aqueous extra- and intracellular fluids, must be affected by drug hydrophobicity. Also, the affinity of a drug towards nonpolar binding sites of blood proteins or a receptor facing an aqueous environment depends on drug hydrophobicity.

What is observed as the hydrophobicity of solutes depends mostly on the environment. However, when comparing behaviour of various solutes in the same environment a quantitative scale can be obtained which reflects differences in abilities of individual solutes to participate in hydrophobic interactions. Such a scale may be constructed by determining the logarithm of the partition coefficient in studies of distribution of the substances between an immiscible polar and nonpolar solvent pair. It is assumed after Collander [25] that partition coefficients for individual solutes may differ if determined in different organic-water solvent systems but their logarithms are linearly related. However, it was observed by Collander himself that the simple relationship became poorer as the polarity differences between the organic solvents in the two aqueous partition systems became larger. Thus, a good correlation can be expected for the 1-octanol-water *versus* the pentanol-water system, while the combination of cyclohexane-water *versus* 1-octanol-water is typical of poor correlation. If there is a unique, one-dimensional hydrophobicity scale, the question arises of which system is more reliable for parametrization of the hydrophobicity of compounds. There is no particular scale and there are as many hydrophobicity parameters as there are partition systems. It can be expected, on the other hand, that hydrophobicity parameters determined in different organic-water systems will be more or less intercorrelated and will reflect the "phobia" of the solutes for the aqueous environment. Thus, it can be said about a given substance that it is hydrophobic; however the degree of hydrophobicity depends on the circumstances.

As the composition of individual partitioning sites in a living organism is not known and there are environmental differences within body compartments, a single hydrophobicity scale is not suitable for characterization of the interactions between drugs and biological systems. The 1-octanol-water partition system introduced by Hansch and co-workers [26,27] is the common reference system for determination of the logarithm of the partition coefficient, $\log P$, widely used in medicinal chemistry. The

1-octanol-water system provides a single, continuous scale for comparing the hydrophobicity of drugs. There is no proof that the 1-octanol-water system is the best possible model for biological permeation barriers and binding sites but large compilations of $\log P$ data provide the basis for numerous satisfactory bioactivity descriptions.

11.3. Chromatographic Determination of Hydrophobicity

11.3.1. General Assumptions

The standard "shake-flask" method of the determination of partition coefficients in the 1-octanol-water or other liquid–liquid systems has several serious disadvantages, the most important being its tediousness and poor interlaboratory reproducibility [12,28]. Instead of measuring partition coefficients by equilibration methods, partition chromatographic data can be used. It has been assumed that the partition thin-layer chromatographic value, R_M, and the logarithm of the partition coefficient, $\log P_s$, determined in a system *identical* with the chromatographic system are related by the Martin equation (29):

$$R_M = \log P_s + \log V_S/V_M, \tag{11.1}$$

where V_S and V_M are the volumes of the stationary and mobile phases, respectively. A slope of exactly unity in Eq. (11.1) could be expected only if the chromatographic phases were identical with the two phases in the classical "shake-flask" experiment, which is practically rather impossible.

11.3.2. Correlations between RP HPLC Retention Parameters and 1-Octanol-Water Partition Coefficients

Certainly, one can expect close correlations between $\log P$ and RP HPLC retention parameters if the chromatographic system closely resembles the conventional 1-octanol-water partition system. Several procedures to make such a system have been reported based on dynamically coating a stationary phase with 1-octanol and using 1-octanol-saturated aqueous eluent [30–32]. The correlations were satisfactory but serious technical problems made the approach impractical.

To get high correlations between retention parameters determined in a stable RP HPLC systems and $\log P$ individual authors recommended a specific treatment of stationary phase before the use and the presence of various additives in the mobile phase [33–36]. Nonetheless, the correlations of such determined chromatographic data with $\log P$ were good only as long as the solutes analysed were more or less closely related (congeneric). According to a very recent paper by Huber and co-workers [37], the proposed method of "determination of true octanol-water partition coefficients by means of solvent generated liquid–liquid chromatography" is devoid of the shortages of former chromatographic methods of determination of hydrophobicity and significantly improves the prediction power and accuracy of the $\log P$ estimation.

As opposed to earlier tendency in recent publications on relationships between $\log P$ and standard reversed-phase liquid chromatographic parameters (both from HPLC and TLC) rather moderate correlations are reported [38–41]. In case of a series of congeneric solutes the correlation between $\log k'_w$ and $\log P$ reported by Clark *et al.* [38], $r = 0.953$, appears reliable and realistic. However, poor correlations have been reported even for congeneric pyrazine derivatives [42] and oxazoline derivatives [43]. On the other hand, perfect correlations ($r = 1.000$) were found between isocratic $\log k'$ and $\log P$ for a structurally diverse series of polymer additives [44].

To retain $\log P$ as the RP HPLC retention descriptor some authors introduce empirical corrections to $\log P$ or the hydrophobic substituent (fragmental) constants [45,46]. Whereas such correlations may be of use for retention prediction they help little in understanding of the mechanism of separations. The same holds true if correlation between $\log k'$ and $\log P$ is improved due to introduction of indicator variables [47,48], molecular refractivity [49] or Hammett sigma values [50] into regression equations. In ref. [49] statistical significance of the refractivity term in the equation reported may be questioned.

Patel *et al.* [51] in their QSRR study employed octanol-water partition coefficient to account simultaneously for changes in solute structure and mobile phase composition. The authors modeled ln k' from RP HPLC by the equations:

$$\ln k' = A + B(\log P/P_{sm}) + C(1/P_{sm}^2), \tag{11.2}$$

or

$$\ln k' = A' + B'(\log P/P_{sm}) + C'(\log P/P_{sm}^2), \tag{11.3}$$

where A, B, C, A', B' and C' are regression coefficients, $\log P$ is logarithm of 1-octanol-water partition coefficient of the solute and P_{sm} is the calculated 1-octanol-water partition coefficient of solvent mixture containing water and either methanol, acetonitrile or tetrahydrofuran. The P_{sm} is calculated from the formula

$$\log P_{sm} = \sum_{i}^{n} (x_i \log P_{s,i}), \tag{11.4}$$

where x_i is the mole fraction of the ith solvent component, $\log P_{s,i}$ is logarithm of 1-octanol-water partition coefficient of the ith solvent component and n is the total number of pure solvents present in the solvent mixture.

The poor performance of $\log P$, determined in a liquid–liquid partition system, in modeling of the RP HPLC retention suggests differences in the two types of partition processes. There are evidences that in case of hydrocarbonaceous stationary phases the solute molecule can vertically penetrate into the bonded hydrocarbon layer [52]. Besides, retention on such phases is affected by the surface density of the bonded alkyl chains [53]. In such a situation chromatographic process cannot directly be modeled by bulk organic-water partitioning processes, because the nonpolar stationary phase is an interphase (immobilized at one end) and not a bulk medium.

In practice, the RP HPLC on chemically bonded nonpolar phases has been commonly employed for hydrophobicity parametrization. Advantages of the RP HPLC method for this purpose are numerous [10,12]. Although there are many reports on nearly perfect correlation (or even identity) of logarithms of capacity factors, $\log k'$, with the "shake-flask" partition data (mostly $\log P$ from the 1-octanol-water system) dissimilarities between the "shake-flask" and chromatographic partition systems are evident. In RP HPLC the partition of the solutes may take place between the mobile phase and a stationary zone formed by preferential adsorption of the organic component of the mobile phase on the "nominal" stationary phase. The solvent composition of the stationary zone is different from that of the mobile phase and varies as a function of distance from the surface of the stationary phase particles [54]. Partitioning of individual solutes may take place in zones with different distances from the solid surface [55].

It is obvious that the RP HPLC systems differ from a regular liquid–liquid partition system if one realizes that in chromatography quasi-chemical equilibria concern the following types of interactions: solute-stationary phase (zone), solute-solvent (*i.e.*, solute-water, solute-organic modifier), solvent-stationary phase (*i.e.*, water-stationary phase, modifier-stationary phase), solvent–solvent (*i.e.*, modifier-modifier, modifier-water, water–water), and in case of some reversed-phase materials the interactions between the flexible fragments of a stationary phase (stationary phase-stationary phase interactions) [10,56]. At higher concentrations of the solute, solute–solute interactions cannot be neglected.

In spite of scientifically justified ambiguities the reversed-phase chromatographic methods of assessment of hydrophobicity of drugs and environmentally important substances have officially been acknowledged and included in the "OECD Guidelines for Testing Chemicals".

From the point of view of environmental sciences the hydrophobicity parameter is one of the most important index of contamination [57]. Correlations between $\log P$ and chromatographic measures of hydrophobicity of environmental pollutants were extensively studied by De Vogt and co-workers [58–61]. As regards drugs and drug candidates the literature till 1986 was reviewed in a previous book by this author [10]. Since that time numerous publications appeared regarding correlations between $\log P$ and chromatographic retention parameters. It would be extremely difficult to critically review all these publications. In this author opinion, there are several papers which especially deserve recommendation because of scientific objectivity and methodological novelty, *e.g.*, refs. [62–70].

11.3.3. Role of Mobile Phase

Assuming the extrathermodynamic linear free-energy relationships, one may consider the logarithm of the HPLC capacity factor, $\log k'$, as reflecting specific structural features of a series of compounds which are responsible for their hydrophobicity. Basically, the advantages of the RP HPLC methods of hydrophobicity quantitation are due to the possibility of using organic modifiers in binary aqueous eluents. However, as discussed above, the presence of organic modifier in mobile phase makes the interactions determining chromatographic separation extremely complex. At first, there is no rationalization for a choice of an individual solvent and its composition which would have been the most suitable for hydrophobicity parametrization in case of given solutes. At a fixed concentration of a given solvent, the solute x may appear more hydrophobic than the solute y, whereas the reverse seems true at another concentration of the eluent. Evidently, different eluent compositions provide different measures of hydrophobicities of the solutes.

In fact, an unequivocal determination of absolute $\log k'$ data at different mobile phase compositions becomes questionable if the uncertainity is realized in the dead volume accompanying the changes in eluent composition [71,72].

Since Boyce and Millborrow [73] applied the retention parameters determined at various organic-water eluent compositions and extrapolated them to the pure water eluent, it became a common practice to employ such extrapolated data as the measures of hydrophobicity. The extrapolation is based on assumption of the linear Soczewiński–Wachtmeister relationship [74] between $\log k'$ and the volume fraction of the organic modifier in a binary aqueous eluent. It has been demonstrated that the Soczewiński–Wachtmeister equation in case of RP HPLC holds at best only at a limited solvent composition range which varies depending on the solute and the chromatographic system employed [75,76]. In effect, the values of the logarithm of capacity factor extrapolated to a pure aqueous eluent (the intercepts in the Soczewiński–Wachtmeister equation denoted commonly by $\log k'_w$) are usually different from those determined experimentally where possible, and depend on the organic modifier employed. Because of this observation some authors [77] are inclined to believe that the extrapolation of capacity factors to 0% organic modifier is a manipulation and the value of $\log k'_w$ itself has no physical meaning [77,78].

Interpretation of the intercept of the Soczewiński–Wachtmeister equation as the logarithm of capacity factor corresponding to pure water (buffer) eluent is certainly misleading. The parameter is not devoid of merits, however, as it may be regarded [12] as a means of normalizing retention because the magnitude of $\log k'_w$ is determined by a change in retention induced by a change in the mobile phase composition, rather than by absolute retention under fixed chromatographic conditions.

If extrapolation to pure water is a normalization of hydrophobicity measures, then the question arises of the most appropriate description of dependence of retention parameters on the composition of mobile phase. The first problem is the range of eluent composition within which the relationship is observed. The linearity of $\log k'$ *versus* percent of organic modifier in the eluent varies with the HPLC system employed and depends specifically on the structure of individual solutes. It has been suggested [79] that the reversed-phase retention mechanism becomes discontinuous at very high and low water concentrations in the mobile phase. However, this discontinuity starts at different mobile phase compositions for individual solutes. Technically, various kinds of linearizations are applied. The most common is plotting of $\log k'$ *versus* volume percent of organic modifier in aqueous eluent (X). However, $\log k'$ *versus* $\log X$ [80] or $\log k'$ *versus* $\log(1/X)$ [81] relationships are also occasionally employed. Lanin and Nikitin [56] derived relationships substantiating the plotting of the reciprocal of capacity factor ($1/k'$) against the mobile fraction of the organic component in binary water-organic mobile phases.

Some authors argue that volume fraction is not the most appropriate parameter for regressing $\log k'$ data and that actual measured solvent polarity should be used [82]. To measure the polarity of organic-water solutions, Reichardt's dye, ET(30), has been used due to its large solvatochromic shift. In view of subsequent reports [83–85] the improvement of the linearity of plots of $\log k'$ *versus* mobile phase composition is not as obvious when using ET(30) parameter instead of volume percent of organic modifier. For a wider composition range of aqueous mixtures of methanol, acetonitrile and tetrahydrofuran the nonlinearity of $\ln k'$ *versus* ET(30) was recently proved [86].

Isocratic capacity factors determined with various organic modifiers depend naturally on the properties of the modifier. One could expect, however, the values extrapolated to pure water, $\log k'_w$, to be independent of the organic modifier used. Unfortunately, this is not usually the case. Thus, different modifiers yield different chromatographic measures of solute hydrophobicity. There is no reason to assume that one modifier provides a better measure of hydrophobicity than another. If the reference hydrophobicity scale is that of 1-octanol-water partition system, then individual organic modifiers of the RP HPLC eluents appear advantageous. Braumann *et al.* [87] strongly advocate the view that a general relationship between $\log P$ and $\log k'_w$ can only be expected to exist for capacity factors determined in methanol-water eluents. According to these authors similar solute-solvent interactions operate in methanol-water and octanol-water system whereas other organic modifiers (acetonitrile, tetrahydrofuran) introduce interactions that are not present in 1-octanol-water system. Although Braumann *et al.* [87] postulate the identity of $\log k'_w$ and $\log P$, there are evidences that even in case of nonpolar solutes, such as chlorobiphenyls and alkylbenzenes, separate regressions of $\log P$ *versus* $\log k'_w$ have to be developed for each class of compounds [78].

Thin-layer chromatographic determination of hydrophobicity deserves a brief separate discussion. Since the Boyce and Millborrow publication [73], the R_M values have often been extrapolated to zero organic modifier concentration in mobile phase. Marked deviations from the linearity have been reported and explained by dual partition/adsorption mechanism of retention. It has been assumed that the sorbent partially retains its adsorptive characteristics, even after impregnation with a hydrophobic substance [88]. It has also been found that the retention parameters depend on organic modifier in aqueous eluent [89]. Limited range of linearity of the R_M *versus* eluent composition has since that time been often reported [90–92].

Interesting aspect of RP TLC is the dependence of R_M on pH. There are reports [93–95] indicating the lack of influence of solvent pH on R_M. Recent results of Mannhold and co-workers [96,97] suggest that the influence of solvent pH on R_M values is negligible in case of low organic modifier content in the mobile phase. With increasing modifier concentration, adsorption becomes more prominent and an influence of pH on R_M is detected in the case of basic solutes.

During last quarter of century Biagi and co-workers studied relationships between R_M and $\log P$ for about 750 drugs and chemicals. The TLC systems introduced and extensively studied by the group consisted of silica layers impregnated with silicone oil and various organic–buffer mobile phases. Tens of papers have been published by Biagi's group. Summary of the research and main conclusions may be found in recent reports [98–102]. According to Biagi *et al.* [101] these are as follows:

(a) The intercepts of the linear relationship between R_M values and concentration of the organic modifier in the aqueous mobile phase can be considered as a measure of the partitioning of compounds between silicone oil and an aqueous buffer.
(b) For the more hydrophilic solutes, which can migrate in a reliable way at 0% organic modifier, very good correlation (with intercept and slope close to 0 and 1, respectively) exist between the experimental and extrapolated R_M values.
(c) The intercepts of the R_M *versus* organic modifier concentration relationships are not dependent on the nature of the organic modifier.
(d) There is a linear relationship between slopes and intercepts of the R_M *versus* organic modifier concentration relationships. Because of these relationships, the use of the slopes of the TLC equations as an alternative lipophilic parameter is valid as suggested also by other authors [103,104]. However, the slope *versus* intercept relationship can be found only when dealing with strictly congeneric compounds. It is rather difficult to define congenericity in terms of chromatographic behaviour. Besides, the variability of the slopes is much smaller than that of intercepts. In such a situation the extrapolated R_M values are recommended as a measure of hydrophobicity.

The impressive scientific efforts of Biagi's group resulted in establishing the RP TLC systems based on silicone oil as a reliable tool for comparing hydrophobicity of drugs. There are some advantages of TLC over HPLC in evaluating hydrophobic properties of solutes, *e.g.*, the TLC requires considerably less laboratory equipment and a number of solutes can be simultaneously analysed. However, the accuracy in TLC demands $0.2 < R_f > 0.8$ [105]. This gives a range of R_M of less than one and half decades. Obviously, TLC is less reproducible than HPLC. As there may be a selective adsorption of mobile phase components in TLC, a quasi-gradient elution may be observed for less retained solutes [106]. The same is true for buffer components. Also, precise control of pH, ionic strength, humidity and temperature during the separation process calls problems in TLC. All this, and additionally, a selection of silicone oil as a rather not typical hydrophobic medium, explains the limited interest in application of Biagi's approach in hydrophobicity studies.

11.3.4. Role of Stationary Phase

In both the solvophobic [107] and solubility parameter [108] theory of RP HPLC, it is assumed that the influence of the stationary phase on the relative retentions of individual solutes is limited. However, hydrophobicity is as much a "phobia" towards aqueous environment as a "philia" towards nonpolar species (lipophilicity). By no means can the chemistry of the contact of the solute with the stationary phase be neglected.

For years the octadecyl-bonded silica (ODS) stationary phases were commonly employed in hydrophobicity studies. Yet, the retention data obtained with nominally the same type of reversed-phase columns under identical mobile phase conditions are hardly comparable [109]. Differences within ODS packing materials may be attributed to surface area of the silica gel, different pore size distribution, nonhomoenergetic silica surfaces [110], metal impurities [111], functionalities of octadecylchlorosilanes coupled to the silanol groups, chemistry of endcapping [112], type and quantity of free silanols, surface area of the bonded phase, carbon content [87], etc. Thus, it is not surprising that differences in retention are observed even when various batches of the same manufacturer's product are used and there is a lack of standardization of the ODS material. In other words, different hydrophobicity measures can be expected when using individual ODS columns.

11.3.4.1. Specific Solute-Stationary Phase Interactions

To get an interlaboratory comparable, universal chromatographic hydrophobicity scale like that of $\log P$, the classical alkyl-bonded silica materials are not suitable. According to Braumann *et al.* [87] capacity factors obtained by extrapolation of retention data from methanol-water eluents to 100% water are insensitive to variations of stationary phase properties. Other authors propose modification of either the mobile or stationary phase to reduce the stationary phase specific effects, like interactions of free silanols. Stadalius *et al.* [113] suggest using a pH between 2.5 and 3.5 with higher buffer concentrations, potassium salts instead of sodium, and the addition of amine modifiers such as triethylamine or dimethyloctylamine. Minick *et al.* [114] propose adding of 1-octanol and 1-decylamine to the eluent composed of methanol and 4-morpholinopropanesulfonic acid buffer. Neither of these approaches appears fully effective, especially when dealing with chemical structures of real meaning for medicinal chemistry, *e.g.*, the nitrogen-comprising heterocyclic derivatives. There is a little chance that special treatments make the RP HPLC systems employing ODS identical with the 1-octanol-water partition system. Thus, the opinion is expressed occasionally by medicinal chemists that chromatographic measures of hydrophobicity are not much reliable. Eadsforth [115] and Leo [116] suggest that if one wishes to predict partition of an individual solute in the 1-octanol-water system then the *a priori* calculated from structural formula $\log P$ values (CLOGP) can often be considered more trustworthy than those determined by HPLC. That does not mean that the chromatographically derived parameters are less trustworthy as descriptors of hydrophobicity.

A serious disadvantage of the silica based reversed-phase materials is their chemical instability at pH above 8. The reference $\log P$ values are determined (or calculated) for neutral, nonionized forms of solutes. Chromatographic determination of hydrophobicity of nonionized forms of organic bases cannot be performed directly on silica based materials. Since biological processes occur in neutral (occasionally acidic) pH hydrophobicity data determined in alkaline environment may be without relevance to pharmacology. As far as the pharmacokinetic processes of membrane permeation are concerned, the apparent hydrophobicity of drugs at the more or less neutral pH is decisive. However, it is assumed that the protein and receptor hydrophobic binding involves the nonionized form of drugs. Thus, hydrophobicity of nonionized form of a drug may affect its overall activity. The affinity of a drug towards proteins or receptors (pharmacodynamics) appears to depend on different hydrophobicity measures than tissue penetration of the drug (pharmacokinetics).

11.3.4.2. New Stationary Phase Materials

In attempts to provide a universal, continuous chromatographic hydrophobicity scale (not necessarily mimicking $\log P$) several new RP HPLC materials have recently been tested [117]. These materials are claimed to be devoid of the major problems of alkyl-bonded silicas, *i.e.*, they have no accessible free silanols and they are chemically stable at a wide pH range.

Poly(styrene-divinylbenzene) copolymers (PS-DVB) are stable over a pH range of 1 to 14. They are reported to provide moderate correlations with $\log P$ which hold usually only within subgroups of congeneric solutes [121,122]. However, columns packed with PS-DVB are characterized by low efficiency and the material suffers from excessive shrinkage and swelling [43,44]. Several polymeric phases having a chemically bonded octadecyl moiety have been tested in hydrophobicity determinations. Such phases like octadecylpolyvinyl copolymer or rigid macroporous polyacrylamide with bonded octadecyls do not undergo swelling nor shrinkage and offer the possibility of having reasonable flow rate without undesirable pressure increase at the column inlet [121,123]. Depending on the specific phase used the reported correlations with $\log P$ of test solutes are low [114] or at best as good as obtained with the octadecylsilica phase [121,124,125]. There are evidences, however, that individual polymeric phases provide specific input to retention. For example, the octadecylpolivinyl copolymer was reported to be less hydrophobic than alkylsilicas but strongly retains specific aromatic compounds [126].

In recent years great progress was achieved in the technology of silica-based reversed-phase materials. Due to high C18 bonding densities, the significant protection of ODS phases against hydrolysis by extremes of pH was attained [127]. Commercially available phases exhibit high level of silanol deactivation [128]. The idea of coating of a chromatographic support material with a layer of a polymer was also adapted to modify the silica-based reversed-phase materials. Othsu *et al.* [129] deposited a silicone monomer on silica and then carried out in situ polymerization and attachment of octadecyl ligands. Similar polymer-coated octadecyl reversed phases were prepared by Hetem *et al.* [130].

Another approach to prepare stable reversed-phase materials with the surface silanols shielded is the encapsulation of silica with a polymeric layer such as polymethylsiloxanes, substituted with long-chain alkyl ligands, polybutadiene [131] and copolymerized vinyl-modified silica with acrylic acid derivatives [132]. Silica-based reversed-phase materials of a new generation provide improved reproducibility of retention due to the reduction of silanophilic interactions. The materials are chemically more stable than regular ODS. However, chromatographic measurements in alkaline conditions must be avoided.

When alumina-based reversed-phase materials appeared there was interest in them from the view point of hydrophobicity parametrization [133]. Alumina is stable over a wide pH range and possesses no interferring silanol groups. The polybutadiene-coated alumina (PBCA) reversed-phase material was introduced by Schomburg and co-workers [134]. Polybutadiene was immobilized on the alumina support surface with the help of a cross-linking reaction that involved radical formation. Due to chemical stability of the PBCA stationary phase, the nonionized forms of acids, bases and neutrals can be analyzed in the same HPLC system operated at an appropriately adjusted pH. Thus, a continuous hydrophobicity scale may be obtained in an easier, faster and more reproducible manner than is the case with the 1-octanol-water system.

More recently a monomeric octadecyl-bonded alumina (ODA) phase was introduced as an HPLC reversed-phase material [135]. The ODA phase has pH stability similar to that of PBCA and exhibits higher chromatographic efficiency. A characteristic feature reported [135] for the monomeric ODA phase is a higher degree of hydrogen bonding solute-stationary phase interactions than on octadecyl-bonded silica.

Carbon supports for RP HPLC [136,137] may also appear interesting for comparative hydrophobicity studies. They have a good chemical stability over a wide pH range, they do not exhibit peak tailing for amines, as do silica-based materials, nor do they adsorb phosphates or carboxylates as do alumina and zirconia polymer-coated phases [138]. On the other hand, their high selectivity could provide information on specific features of the hydrophobicity of the solutes. This information could be of value for structure–activity studies and thus the reported [118] lack of correlation of $\log k'$ determined on graphitic carbon with $\log P$ should not be discouraging.

The RP HPLC approach has been relatively successful in duplicating $\log P$ data derived by traditional techniques. However, if the reason to use the chromatographic system is to model processes in the biophase, then the components of the chromatographic and the biological systems must be comparable. Thus, an RP HPLC system which is used to model transport through biological membranes should be composed of an aqueous phase and an organized phospholipid layer.

Miyake *et al.* [139] derived HPLC hydrophobicity parameters employing a column of silica gel coated physically with dipalmitoyl phosphatidylcholine (DPPC). Seydel *et al.* [140] proposed an HPLC technique using phosphatidylcholine or phosphatidylserine covered infusorial earth as the stationary phase and phosphate buffer as mobile phase to determine partitioning into artificial phospholipids. Leaving aside inconveniences regarding their preparation and stability, the systems with phosphatidyl derivatives adsorbed onto silica most probably do not emulate the lipid dynamics of biological membranes, because the adsorbed lipids are not organized in a manner similar to natural (or artificial) membranes.

Recently, a new RP HPLC stationary phase material, the immobilized artificial membrane stationary phase (IAM), became available (Fig. 11.1) [141]. This phase models natural membranes which are composed of lipids with a polar headgroup and two nonpolar chains. In the IAM phase a lipid molecule, lecithin, is covalently bound to propylamine silica forming confluent monolayers of immobilized membrane lipids. In one form of the IAM, unreacted propylamine moieties are end-capped with methylglycolate. Only one of the alkyl chains is linked to the propylamine silica surface, and the immobilized lipid headgroups protrude away from the stationary phase surface. These charged moieties are the first contact site between solutes and IAM.

Preliminary studies reported encouraging correlations between HPLC retention parameters determined on an IAM and human skin permeation data for a short

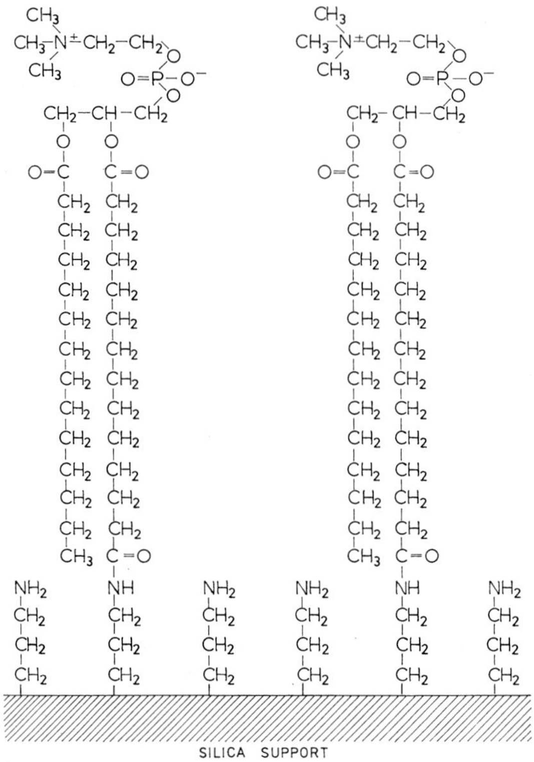

Figure 11.1. Chemical structure of the immobilized artificial membrane (IAM) stationary phase.

series of alcohols and steroids [141,142]. The same biological data showed poor correlation with HPLC retention parameters determined on a regular hydrocarbonaceous reversed-phase HPLC columns.

Correlations between the biochromatographic parameter of hydrophobicity determined on IAM and reference hydrophobicity data were studied for groups of phenothiazine neuroleptics and antihistamine drugs [143], cardiovascular β-adrenolytics, α-adrenomimetics and phenothiazine neuroleptics [144] as well as for a series of steroid hormones and phenol antiseptics [145]. Correlations varied depending on the class of agents studied.

The data considered in the case of β-adrenolytic drugs [144] are collected in Table 11.1. The relationship between the logarithm of the capacity factor determined on the IAM with acetonitrile-0.1M phosphate buffer pH 7.0 10:90%(v/v), $\log k'_{\mathrm{IAM}}$, and the 1-octanol-water partition coefficient corrected for ionization at pH 7.0, $\log D$, is described by the following equation:

$$\log D = 2.055(\pm 0.206)\log k'_{\mathrm{IAM}} - 1.560(\pm 0.123); \qquad (11.5)$$

$$n = 10; \quad r = 0.962; \quad s = 0.32;$$

Table 11.1 Hydrophobicity parameters and pK_a data for a series of β-adrenolytic drugs. (Reprinted with permission from Kaliszan, R., Nasal, A. and Buciński, A., *Eur. J. Med. Chem.* 29, 163. Copyright Publications Elsevier, Paris.)

Drug	pK_a[a]	*log P*[b]	*log D*[c]	*CLOGD*[d]	*CLOGD*[e]	$\log k'_{IAM}$[f]
Acebutolol	9.20	1.77	−0.43	1.61	−0.59	0.528
Alprenolol	9.65	3.10	0.45	2.59	−0.06	0.853
Atenolol	9.60	0.16	−2.44	−0.11	−2.71	−0.301
Betaxolol				2.17		0.929
Bisoprolol				1.69		0.574
Bopindolol				4.86		0.376
Carteolol		−0.46		1.17		0.301
Celiprolol				1.66		0.653
Cicloprolol				2.39		0.813
Dilevalol	9.45	3.09		2.18		1.211
Esmolol				1.53		0.574
Metoprolol	9.70	1.88	−0.82	1.20	−1.50	0.352
Nadolol	9.39	0.71	−1.68	0.23	−2.16	0.176
Nebivalol				3.50		
Nifenalol	8.78			1.16	−0.63	0.176
Oxprenolol	9.50	2.18	−0.32	1.62	−0.88	0.512
Pindolol	9.70	1.75	−0.95	1.65	−1.05	0.512
Practolol	9.50	0.79	−1.71	0.78	−1.72	−0.204
Propranolol	9.45	3.56	1.11	2.75	0.30	1.279
Sotalol	9.05	0.24	−1.81	0.23	−1.82	−0.301
Timolol						0.301

[a] Acidity constant; [b] logarithm of *n*-octanol/water partition coefficient for neutral forms determined experimentally by the shake-flask method; [c] logarithm of experimental *n*-octanol/water partition coefficient corrected for ionization at pH 7.0; [d] logarithm of theoretical *n*-octanol/water partition coefficient for neutral form of solute calculated by fragmental method; [e] logarithm of theoretical *n*-octanol/water partition coefficient corrected for ionization at pH 7.0; [f] logarithm of HPLC capacity factor determined on an immobilized artificial membrane column; [a,b,d] data taken from Craig [146]; [c,e,f] data calculated or determined in ref. [144].

where n denotes the number of solutes considered to derive the regression equation, r is the correlation coefficient and s is the standard error of estimate; numbers in parentheses account for the standard error of the regression coefficients.

There is a lower correlation between $\log k'_{\mathrm{IAM}}$ and the theoretically calculated from fragmental constants 1-octanol-water hydrophobicity parameter [146] corrected for ionization at pH 7.0, CLOGD:

$$\mathrm{CLOGD} = 1.620(\pm 0.302)\log k'_{\mathrm{IAM}} - 1.693(\pm 0.172); \tag{11.6}$$

$$n = 11; \quad r = 0.873; \quad s = 0.47.$$

Most probably the lower statistical quality of Eq. (11.6) with regard to Eq. (11.5) reflects the inadequacies of CLOGD.

For seven members of the set of psychotropic drugs (Table 11.2) [144] chromatographed on the IAM, both $\log P$ and pK_a data were found and the $\log D$ values could be calculated which corresponded to pH 7.0 of the HPLC eluent. The following equation related the two hydrophobicity parameters:

Table 11.2 pK_a values, hydrophobicity parameters and antihaemolytic activity of a series of phenothiazine neuroleptics. (Reprinted with permission from Kaliszan, R., Nasal, A. and Buciński, A., *Eur. J. Med. Chem.* 29, 163. Copyright Publications Elsevier, Paris.)

Drug	pK_a [a]	*log P* [b]	*log D* [c]	*CLOGD* [d]	*CLOGP* [e]	$\log k'_{IAM}$ [f]	log 1/*C* [g]
Chlorpromazine	9.30	5.35	3.05	5.20	2.90	1.440	4.72
Fluphenazine	8.10	4.36	3.23	5.90	4.77	1.401	5.00
Perphenazine	7.80	4.20	3.34	5.57	4.71	1.373	4.82
Prochloperazine	8.10			6.15	5.02	1.782	5.22
Promazine	9.40	4.55	2.15	4.28	1.88	1.171	4.55
Promethazine	9.10			4.65	2.55	1.195	
Propiomazine	9.10			5.00	2.90	1.265	
Thioridazine	9.50	5.90	3.40	6.42	3.92	1.747	5.30
Trifluoperazine	8.10	5.03	3.90	6.48	5.35	1.754	5.22
Trifluopromazine	9.20	5.19	2.99	5.53	3.33	1.474	4.96

[a] Acidity constant; [b] logarithm of experimental *n*-octanol/water partition coefficient for neutral forms determined by the shake-flask method; [c] logarithm of experimental *n*-octanol/water partition coefficient corrected for ionization at pH 7.0; [d] logarithm of theoretical *n*-octanol/water partition coefficient for neutral form of solute calculated by fragmental method; [e] logarithm of theoretical *n*-octanol/water partition coefficient corrected for ionization at pH 7.0; [f] logarithm of HPLC capacity factor determined on an immobilized artificial membrane column; [g] logarithm of reciprocal of haemolytic concentration; [a,b,d] data taken from Craig [146]; [c,e,f] data calculated or determined by Kaliszan *et al.* [144].

$$\log D = 2.142(\pm 0.621)\log k'_{\mathrm{IAM}} - 0.019(\pm 0.927) \tag{11.7}$$

$$n = 7; \quad r = 0.839; \quad s = 0.32.$$

Note that slopes are similar in Eqs. (11.5) and (11.7). In case of nonionized steroids the correlation between $\log k'_{\mathrm{IAM}}$ and the classical hydrophobicity parameter, $\log P$, was $r = 0.911$. On the other hand, there was no significant correlation between these quantities in the case of phenolic derivatives. Also Ong *et al.* [147] recently reported moderate correlations between $\log P$ and $\log k'_{\mathrm{IAM}}$.

Certainly, $\log k'_{\mathrm{IAM}}$ from IAM columns reflects different characteristics of hydrophobicity than the reference $\log P$. In such a situation $\log k'_{\mathrm{IAM}}$ is an interesting alternative to $\log P$ as a predictor of bioactivity. The more so because, contrary to $\log P$, the $\log k'_{\mathrm{IAM}}$ is derived in a simple, fast and reproducible manner.

Leaving aside the problem of correlation between $\log k'_{\mathrm{IAM}}$ and $\log P$ there are increasing evidences that the chromatographically determined hydrophobicity parameter may better than $\log P$ predict drug permeation through biological membranes [148,149] and binding to biomacromolecules [150,151].

11.3.5. Search for a Continuous Chromatographic Hydrophobicity Scale

The $\log P$ hydrophobicity scale may not be the best one but it has a strong advantage: it is continuous. Attempts to construct a continuous chromatographic hydrophobicity scale comprising both acidic and basic solutes were hampered by the instability of the ODS material at alkaline pH and silanophilic interactions. With the new generation of the RP HPLC stationary phases available a convenient, versatile chromatographic method of hydrophobicity parametrization could be elaborated which would be applicable to a wide range of chemical structures. Although the chromatographic hydrophobicity scale need not to duplicate nor to be perfectly parallel to the log *P* scale, nevertheless it should quantitatively reflect the relative "phobia" of the solutes towards aqueous environment. The polybutadiene-coated alumina stationary phase (PBCA) appeared particularly suitable for the purpose due to its chemical stability, lack of silanols and good mechanical properties. Employing the PBCA phase the studies were undertaken in this author's laboratory to construct a continuous chromatographic hydrophobicity scale, to relate it to the classical log *P* and to apply the hydrophobicity data for deriving of the quantitative structure-pharmacological activity relationships (QSAR).

With the PBCA stationary phase it was possible to determine the HPLC capacity factors at both acidic (pH 1.65) and alkaline (pH 10.7) conditions. The set of test solutes (Table 11.3) comprised diverse structures – mostly organic bases but some acids and neutrals included – of a wide range of hydrophobicity as expressed by the log *P* data. The log *P* data were taken as means of the values reported by Hansch and Leo [152] for the nonionized forms of the solutes studied. The relationship between log *P* and log *k′* presented in Fig. 11.2 is described by the equation

$$\log P + 1.49 \log k' + 2.36. \tag{11.8}$$

Table 11.3 Logarithms of RP HPLC capacity factors (log *k′*) of nonionized forms of basic, neutral and acidic solutes determined on polybutadiene-coated alumina with methanol–buffer 50:50 %(v/v) eluent of adjusted pH and ionic strength and literature [152] data of log *P*. (Reprinted with permission from Kaliszan, R., Blain, R.W. and Hartwick, R.A., *Chromatographia* 25, 3, 1988. Copyright Friedr. Vieweg & Sohn.)

No.	*Compound*	*log P*	*log k′*
1	Antipyrine	0.23	–
2	4-Chloropyridine	1.28	−0.689
3	Acridine	3.40	0.410
4	Procaine	1.90	−0.643
5	(+) Ephedrine	1.02	−0.666
6	4-Chlorobenzoic acid	2.65	0.111
7	4-Chloroaniline	1.64	−0.099
8	*p*-Toluidine	1.42	−0.513
9	Benzamide	0.65	−1.546
10	Atropine	1.81	−0.332
11	9-Aminoacridine	2.74	0.038
12	Sulfanilamide	−0.72	–
13	Benzoic acid	1.95	−0.401
14	Acetanilide	1.16	−0.990
15	Phenol	1.48	−0.582
16	Chlorobenzene	2.83	0.632
17	Caffeine	−0.07	−1.142
18	Aniline	1.08	−0.865
19	Diphenylamine	3.44	0.927
20	N-Phenylanthranilic acid	4.36	0.960
21	Acetophenone	1.66	−0.382
22	Phenothiazine	3.78	1.223
23	Biphenyl	4.06	1.318
24	Chloropromazine	5.35	1.638

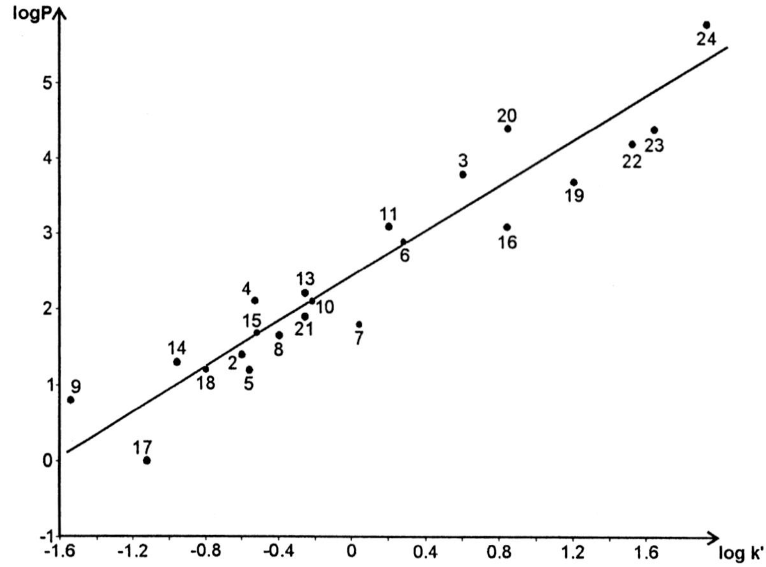

Figure 11.2. Relationship between logarithms of isocratic capacity factors (log *k*′) and logarithms of 1-octanol-water partition coefficients (log *P*) for nonionized forms of test solutes listed in Table 11.3. (Reprinted with permission from Kaliszan, R., Blain, R.W. and Hartwick, R.A., *Chromatographia* 25, 3, 1988. Copyright Friedr. Vieweg & Sohn.)

Equation (11.8) is highly significant ($p < 10^{-11}$) and is characterized by the correlation coefficient $r = 0.96$. The equation comprises the chemically unrelated solutes differing in 1-octanol-water partition coefficient by five orders of magnitude. Certainly, there is a scattering of data in Fig. 11.2. This results in part due to uncertainities in log *P* data and is comparable to the scattering of data points reported for similar relationships derived for limited sets of much less diverse structures when employing standard reversed-phase columns.

Chromatographic capacity factors determined on PBCA at pH 11.5 for a series of nitrogen-comprising organic bases were demonstrated to correlate well with log *P* as opposed to analogous data determined at neutral pH [153].

Direct comparison of performances of PBCA and ODS phases in parametrization of hydrophobicity of drugs was done for a series of 16 imidazole derivative drugs (Fig. 11.3).

The PBCA column was operated at pH 10.9 of the buffer. With the ODS column pH was 7.0. Methanol–buffer mobile phases in seven proportions of the components were employed. In case of the ODS system the approach was applied as proposed by Minick *et al.* [83,154] to limit silanophilic interactions of basic solutes. The approach consists on adding an amine modifier and 1-octanol to the mobile phases.

Advantages of PBCA over ODS for deriving reproducible chromatographic measures of hydrophobicity are typically illustrated in Fig. 11.4 [155]. The PBCA column provides good linearity of log *k*′ *versus* volume percent of methanol at a wide eluent composition range. In case of ODS linearity is at best limited to low (<60% (v/v)) concentration of methanol. The approach by Minick *et al.* [83,154] seems to be ineffective for the organic nitrogen compounds studied. The intercepts of the log *k*′

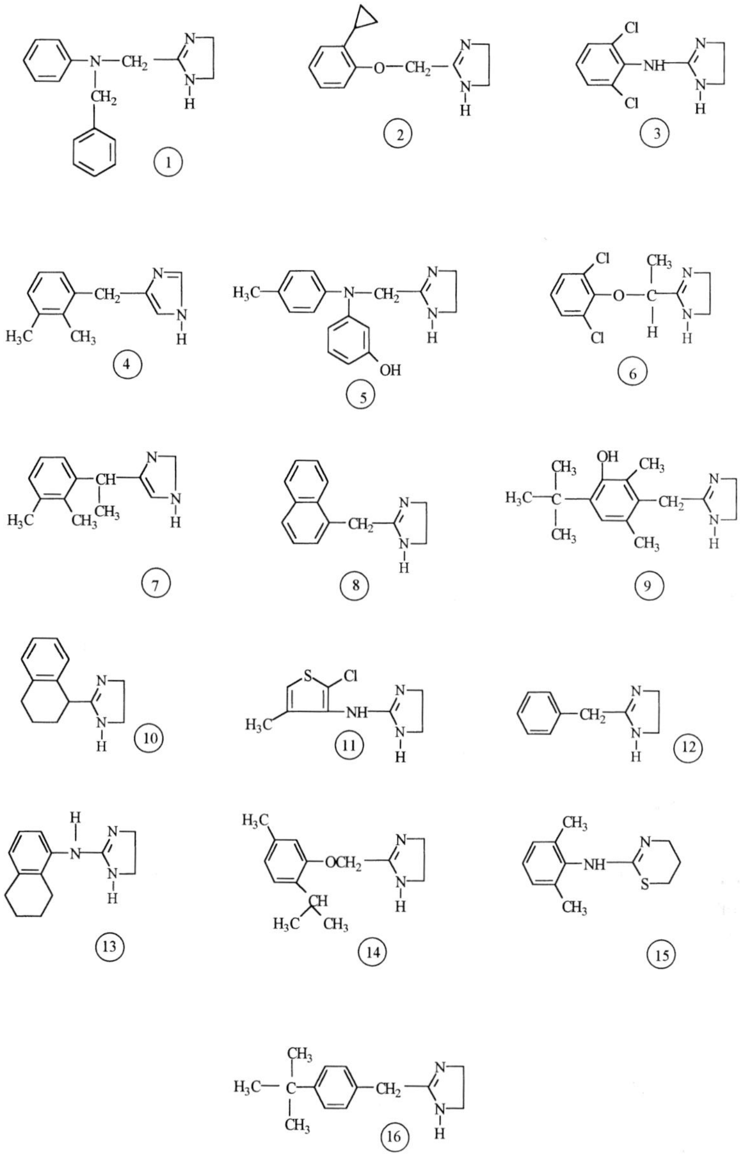

Figure 11.3. Azole derivative circulatory drugs selected to compare performance of polybutadiene-coated alumina and octadecylsilica in hydrophobicity parametrization and to test the applicability of chromatographic data for pharmacological classification of solutes.

versus methanol content for individual solutes, *i.e.*, $\log k'_w$, determined in the two RP HPLC systems correlate poorly ($r = 0.606$, $p < 0.05$). Stronger retentive properties of ODS than PBCA are observed. Due to the lack of silanophilic interactions and the possibility of suppression of solutes ionization, the PBCA reversed-phase material appears especially suitable for parametrization of hydrophobicity of organic bases.

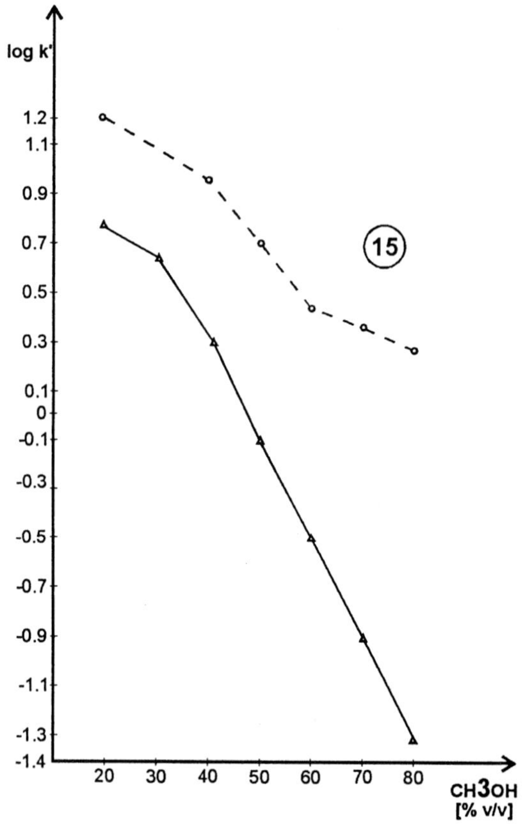

Figure 11.4. Exemplary plot of $\log k'$ against volume fraction of methanol in mobile phase for solute No. 15 from Table 11.3. Solid line corresponds to $\log k'$ data determined on the PBCA column (pH 10.9); broken line corresponds to data obtained on the ODS column (pH 7.0). (Reprinted with permission from Gami-Yilinkou, R. and Kaliszan, R., *Chromatographia* 30, 277, 1990. Copyright Friedr. Vieweg & Sohn.)

Regular behaviour observed for an individual family of solutes in a given HPLC system need not be followed by other groups of compounds. Abnormal retention effects due to interactions with the support material (silica, alumina) may manifest themselves differently for various solute classes and chromatographic conditions. To get better insight into the nature of chromatographic measures of hydrophobicity different solutes and different HPLC systems must be studied. Rather nontypical group of heterocyclic solutes comprising nitrogen but showing acidic character are pyrazine CH- and NH-acids (Fig. 11.5). Three commercially available columns were used to study their retention. One was PBCA-based (Unisphere[R]-PBD, Biotage, Charlottesville, VA, USA). The two were ODS-based.

The standard ODS column produced $\log k'$ values more or less linearly dependent on methanol content but only up to its concentration of 65% (v/v) (Fig. 11.6). For majority of solutes the data obtained at higher methanol content deviated from linearity. The Suplex pKb-100 column, operated both at pH 2.4 and 7.4, yields $\log k'$ data dependent

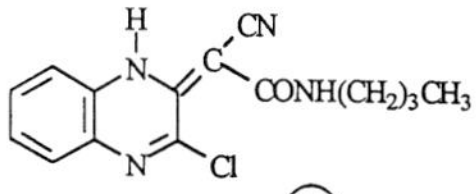

(91)

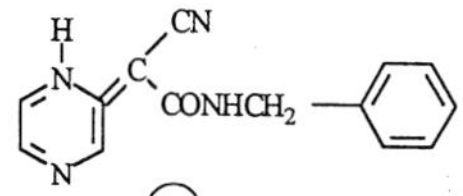

(92)

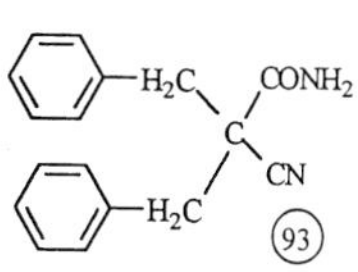

(93)

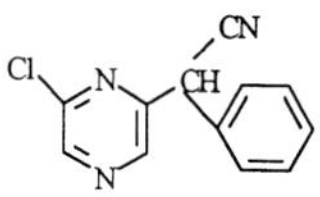

(94)

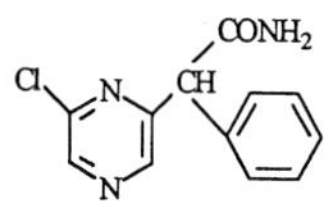

(95)

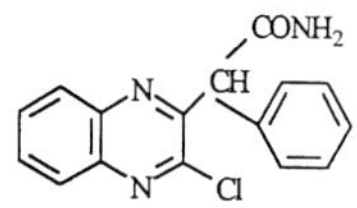

(96)

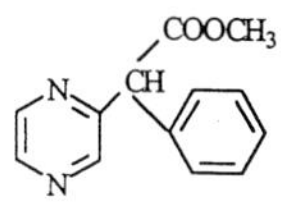

(97)

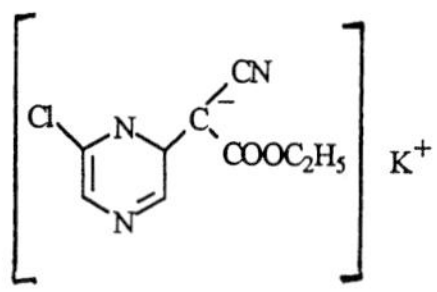

(98)

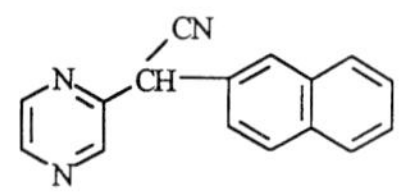

(99)

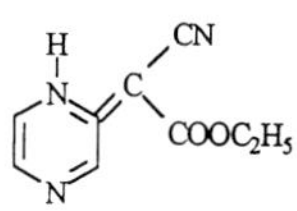

(100)

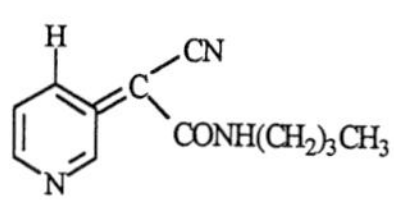

(101)

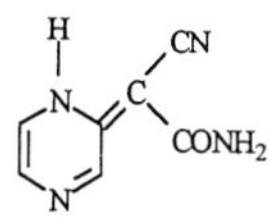

(102)

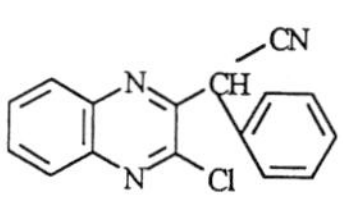

(104)

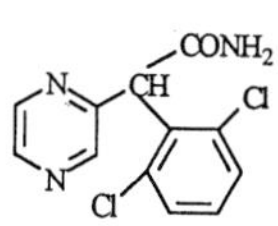

(105)

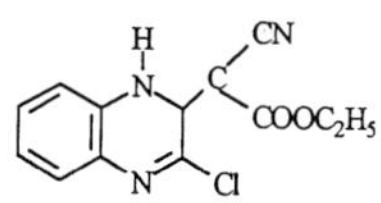

(106)

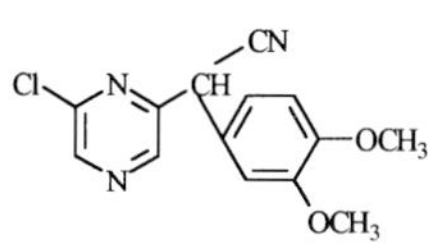

(108)

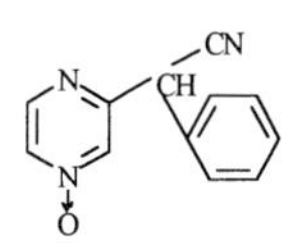

(109)

Figure 11.5. A series of test solutes comprising pyrazine CH- and NH-acids and compound 93, analyzed in seven diverse RP HPLC systems.

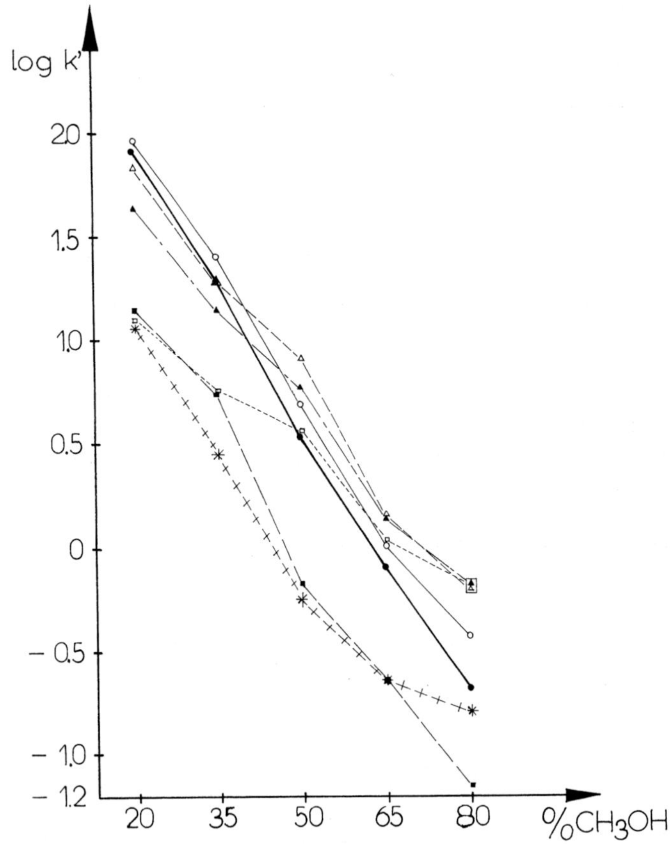

Figure 11.6. Plots of logarithms of capacity factor, log k', against volume percent of methanol in mobile phases in seven HPLC systems for solute 99 (Fig. 11.5) ▲—▲ ODS, pH 2.4; Δ—Δ ODS, pH 7.4; □—□ ODS, pH 7.4, 1-octanol, 1-decylamine; ●—● Suplex pKb-100, pH 2.4; ○—○ Suplex pKb-100, pH 7.4; ■—■ Unisphere-PBD, pH 2.4; ✻—✻ Unisphere-PBD, pH 11.5. (Reprinted with permission from Gami-Yilinkou, R. and Kaliszan, R., *Pol. J. Pharmacol. Pharm.* 44, 515, 1992.)

linearly on volume fraction of methanol within the whole range of mobile phase composition studied, *i.e.*, from 20% to 80% (v/v) of methanol. In case of the Unisphere-PBD column retention at pH 2.4 is higher than at pH 11.5. In several instances, *e.g.*, for solute 106, tenfold difference between respective k' data was noted [156,157].

It appears that adding a higher aliphatic amine to the eluent to suppress the specific interactions of solutes with ODS does not improve column performance in case of the chemical class studied. On the other hand, the masking agents introduce additional variable to the RP HPLC system. Similar lack of advantages of chromatographic measures of hydrophobicity determined in the presence of a masking agent was reported by Bechalany *et al.* [121,124].

The intercepts (log k'_w) of the equations relating log k' to volume fraction of methanol in the mobile phase, determined for individual solutes in the seven RP HPLC systems studied, provide quite independent measures of hydrophobicity. In Table 11.4

Table 11.4 Intercorrelations among the logk'_w data determined for compounds presented in Fig. 11.5 in the following RP HPLC systems: ODS, pH 2.4 (I); ODS, pH 7.4 (II); ODS, pH 7.4, 1-octanol, 1-decylamine (III); Suplex pKb-100, pH 2.4 (IV); Suplex pKb-100, pH 7.4 (V); Unisphere-PBD, pH 2.4 (VI); Unisphere-PBD, pH 11.5 (VII).

	I	*II*	*III*	*IV*	*V*	*VI*	*VII*
I	1	0.6174	0.3541	0.7556	0.6591	0.8295	0.5190
II		1	0.4406	0.6070	0.6983	0.7791	0.8888
III			1	0.4620	0.7288	0.4209	0.2510
IV				1	0.6601	0.8955	0.5828
V					1	0.7011	0.4957
VI						1	0.7853

correlation matrix is given for the $\log k'_w$ data. Intercorrelations among individual HPLC hydrophobicity scales can by no means be called high. The differences in $\log k'_w$ determined in various systems reflect quantitatively differences in properties of solutes and stationary phases.

For the group of pyrazine CH- and NH-acids the specially deactivated ODS columns seem to be the most suitable for assessment of relative hydrophobicities. However, their instability at alkaline conditions precludes the comparison of hydrophobicity of nonionized basic solutes. The PBCA phases appear promising with regard to obtaining an universal, continuous chromatographic scale of hydrophobicity providing that they are deactivated and their retentive properties increased.

11.4. Centrifugal Partition Chromatography

Several authors studied centrifugal partition chromatography (CPC) as a novel technique for measuring liquid–liquid partition coefficients [158–164].

In CPC two nonmiscible liquids constitute the two chromatographic phases: stationary and mobile ones. No solid support for the stationary phase is required. The liquid stationary phase is maintained in a series of mutually connected columns by a centrifugal force. The mobile phase is pumped through the stationary phase. Various CPC systems have been developed, such as the flow-through multilayer coil planet centrifuge [165,166], the horizontal flow-through multilayer coil planet centrifuge [167,168], the toroidal coil planet centrifuge [169,170] and the multichannel cartridges CPC [171,172].

There are the features of CPC which justify interest in this method from the viewpoint of hydrophobicity determination. The retention in CPC is not disturbed by interactions originating from the stationary phase material like silica. Thus in CPC the retention parameters reflect partitioning phenomena similarly like in the "shake-flask" methods of hydrophobicity determination. Also there is no pH limit.

Detailed examination of the applicability of CPC for measuring of partition coefficients of various benzene derivatives was done by El Tayar *et al.* [162]. Using the solvent system hexanol-aqueous buffer (pH 7.4) these authors obtained satisfactory correlation of the equilibrium partition coefficient determined by CPC, $\log P_{CPC}$, with the octanol-water partition coefficient determined by the classical "shake-flask" procedure, $\log P$. Correlation coefficient for the regression was $r = 0.968$. The relationship comprised solutes possessing various functional groups. El Tayar *et al.* [162] concluded that CPC is a method suitable for studying partitioning of organic solutes in a variety of solvents. However, practical problems limit the range of partition coefficients which can be measured by CPC. These problems are duration of elution and the band broadening in the case of more strongly retained solutes. As observed by the same authors, increasing the speed of rotation of the chromatograph improves resolution but at the same time leads to unacceptably high back pressures. A separate problem may be the choice of proper solvents as either the stationary phase or mobile phase. The authors [162] were unsuccesful in their attempts to apply 1-octanol-aqueous buffer system for CPC because of high viscosity of 1-octanol. Latter on Gluck and Martin [173] applied the 1-octanol-water system in CPC and were able to determine $\log P$ over the range of -0.5 to 2.5 log units. El Tayar *et al.* [174] compared the value of four CPC systems for measuring $\log P$. They got excellent correlation with published $\log P$ values obtained by the "shake-flask" method for the $\log P$ range from -3 to $+3$.

Generally, physical properties of solvents limit the reliable hydrophobicity measuring range by CPC. There are problems with determination of the dead volumes as continuous decrease of the stationary phase volume ("bleeding") occurs in spite of presaturation of the phases [175]. One has to note that for CPC a special centrifugal chromatograph is required.

CPC is based on the same principle like the classical countercurrent chromatography (CCC). The difference is that due to the application of centrifugal force, CPC is much faster than CCC. Gago *et al.* [176,177] reported the use of the droplet CCC for hydrophobicity determination of a short series of aromatic compounds. The results obtained and the characteristics of the procedure do not prove real advantages of the method. It must be noted here that using the method of co-current CCC Berthod and co-workers [178,179] succeeded in extending the measurable $\log P$ range to 4.3 units.

11.5. Concluding Remarks

Much effort was devoted to produce convenient chromatographic hydrophobicity measures. As far as the commonly recognized octanol-water reference hydrophobicity scale ($\log P$) is concerned the attempts to reproduce it by means of HPLC have succeeded only partially. Better chance of mimicking the $\log P$ provide centrifugal countercurrent liquid–liquid partitioning systems but inconveniences of these systems hinder their wider application. Still the RP HPLC is the method of choice for parametrization of relative hydrophobicity within series of related solutes. To get appropriate data a basic knowledge is required about the factors affecting chromatographic measures of hydrophobicity. With the new stationary phase materials presently available it is possible to get the HPLC hydrophobicity scales possessing the advantages of the $\log P$ scale. Such scales are continuous and universal, *i.e.*, allowing for analysis of nonionized forms of acidic, neutral and basic solutes. Neither the $\log P$ nor any of the chromatographic hydrophobicity scales is unique, pharmacologically distinguished. Thus, instead of striving for a single "best" hydrophobicity scale all the information on structure of solutes should be exploited which is provided by reproducible chromatographic measurements. By applying chemometric approaches systematic information may be extracted from diversified retention data.

References

1. Overton, E., *Z. Physikal. Chem.* **22**, 189 (1897).
2. Meyer, H., *Arch. Exp. Pathol. Pharmakol.* **42**, 109 (1899).
3. Baum, F., *Arch. Exp. Pathol. Pharmakol.* **42**, 119 (1899).
4. Martin, A.J.P., *Biochem. Soc. Symp.* **3**, 4 (1950).
5. Hansch, C. and Fujita, T., *J. Am. Chem. Soc.* **86**, 1616 (1964).
6. Boyce, C.B.C. and Millborrow, B.V., *Nature* (*London*), **208**, 537 (1965).
7. Biagi, G.L., *Antibitica* **5**, 198 (1967).
8. Huber, J.F.K., Meijers, C.A.M. and Hulsman J.A.R.J., *Anal. Chem.* **44**, 111 (1972).
9. Tomlinson, E., *J. Chromatogr.* **113**, 1 (1975).
10. Kaliszan, R., "*Quantitative Structure-Chromatographic Retention Relationships*", Wiley, New York, 1987.
11. Hafkenscheid, T.L. and Tomlinson, E., *Adv. Chromatogr.* **25**, 1 (1986).
12. Braumann, T., *J. Chromatogr.* **373**, 191 (1986).
13. Terada, T. *Quant. Struct.-Act. Relat.* **5**, 81 (1986).
14. Valko, K., *Trends Anal. Chem.* **6**, 214 (1987).
15. Kaliszan, R., in: "*High-Performance Liquid Chromatography*", Brown, P.R. and Hartwick, R.A. (Eds.), Wiley, New York, 1989; pp. 563–599.
16. Kaliszan, R., *Quant. Struct.-Act. Relat.* **9**, 83 (1990).
17. Kaliszan, R., *Anal. Chem.* **64**, 619A (1992).
18. Kaliszan, R., *Adv. Chromatogr.* **33**, 147 (1993).
19. Kaliszan, R., *J. Chromatogr. A.* **656**, 417 (1993).
20. Lambert, W.J., *J. Chromatogr. A* **656**, 469 (1993).
21. Dorsey, J.G. and Khaledi, M.G., *J. Chromatogr. A* **656**, 485 (1993).
22. Dill, K.A., *Science* **250**, 297 (1990).
23. Ben-Naim, A. "*Hydrophobic Interactions*", Plenum Press, New York, 1980.
24. Kaliszan, R., in: "*High Performance Liquid Chromatography*", Brown, P.R. and Hartwick, R.A. (Eds.), Wiley, New York, 1989; Chapter 4.
25. Collander, R., *Acta Chem. Scand.* **5**, 774 (1951).
26. Hansch, C. and Fujita, T., *J. Amer. Chem. Soc.* **86**, 1616 (1964).
27. Hansch, C. and Leo, A., "*Substituent Constants for Correlation Analysis in Chemistry and Biology*", Wiley, New York, 1979.
28. Kaliszan, R., *Crit. Rev. Anal. Chem.* **16**, 323 (1986).
29. Martin, A.J.P., *Biochem. Soc. Symp.* **3**, 4 (1950).
30. Henry, D., Block, J.H., Anderson, J.L. and Carlson, G.R., *J. Med. Chem.* **19**, 619 (1976).
31. Unger, S.H. and Chiang, G.H., *J. Med. Chem.* **24**, 262 (1981).
32. Lewis, S.J., Mirrlees, M.S. and Taylor, P.J., *Quant. Struct.-Act. Relat.* **2**, 1 (1983).
33. Garst, J.E., *J. Pharm. Sci.* **73**, 1616 (1984).
34. Sabatka, J.J., Minick, D.J., Shumaker, T.K., Hodgson, G.L., Jr. and Brent, D.A., *J. Chromatogr.* **384**, 349 (1987).
35. Gago, F., Alvarez-Builla, J. and Elguero, J., *J. Liquid Chromatogr.* **10**, 1031 (1987).
36. Pietrogrande, M.C., Dondi, F., Blo, G., Borea, P.A. and Bighi, C., *J. Liq. Chromatogr.* **10**, 1065 (1987).
37. Cichna, M., Markl, P. and Huber, J.F.K., *J. Pharm. Biomed. Anal.* **13**, 339 (1995).
38. Clark, C.R., Barksdale, J.M., Mayfield, C.A., Ravis, W.R. and DeRuiter, J., *J. Chromatogr. Sci.* **28**, 83 (1990).
39. Dross, K.P., Mannhold, R. and Rekker, R.F., *Quant. Struct.-Act. Relat.* **11**, 36 (1992).
40. Biagi, G.L., Recanatini, M., Barbaro, A.M., Guerra, M.C., Sapone, A., Borea, P.A. and Pietrogrande, M.C., in: "*QSAR: Rational Approaches to the Design of Bioactive Compounds*", Silipo, C. and Vittoria, A. (Eds.), Elsevier, Amsterdam, 1991, p. 83.
41. Tanaka, N., Tanigawa, T., Kimata, K., Hosoya, K. and Araki, T., *J. Chromatogr.* **549**, 29 (1991).
42. Yamagami, C. and Takao, N., *Chem. Pharm. Bull.* **39**, 1217 (1991).

43. Demotes-Mainard, F., Jarry, C., Thomas, J. and Dallet, P., *J. Liq. Chromatogr.* **14**, 795 (1991).
44. Jinno, K. and Yokoyama, Y., *J. Chromatogr.* **550**, 325 (1991).
45. Haky, J.E. and Young, A.M., *J. Liq. Chromatogr.* **7**, 675 (1984).
46. Petrauskas, A.A. and Svedas, V.K., *J. Chromatogr.* **585**, 3 (1991).
47. Yamagami, C. and Takao, N., *Chem. Pharm. Bull.* **39**, 1217 (1991).
48. Yamagami, C., Yokota, M. and Takao, N., *J. Chromatogr. A* **662**, 49 (1994).
49. Pereira, A.L., Barreiro, E.J.L., Freitas, A.C.C., Correa, C.J.C. and Gomes, L.N.L.F., *J. Liq. Chromatogr.* **14**, 1161 (1991).
50. Tipker, J., Groen, C.P., Van den Bergh-Swart, J.K. and Van den Berg, J.H.M., *J. Chromatogr.* **452**, 227 (1988).
51. Patel, H.B., King, D.N. and Jefferies, T.M., *J. Chromatogr.* **555**, 21 (1991).
52. Tchapla, A., Colin, H. and Guiochon, G., *Anal. Chem.* **56**, 621 (1984).
53. Sentell, K.B. and Dorsey, J.G., *Anal. Chem.* **61**, 930 (1989).
54. Knox, J.H., Kaliszan, R. and Kennedy, G.J., *J. Chem. Soc., Faraday Symp.* **15**, 113 (1980).
55. Men, Y.-D. and Marshall, D.B., *Anal. Chem.* **62**, 2606 (1990).
56. Lanin, S.N. and Nikitin, Y.S., *J. Chromatogr.* **537**, 33 (1991).
57. Noble, A., *J. Chromatogr.* **642**, 3 (1993).
58. De Vogt, P. and Govers, H., *Chemosphere* **15**, 1467 (1986).
59. De Vogt, P., Van Zijl, G.A., Govers, H. and Brinkman, U.A.T., *J. Planar Chromatogr.* **3**, 24 (1990).
60. De Vogt, P., Wegener, J.W.M., Klamer, J.C., Van Zijl, G.A. and Govers, H., *Biomed. Environ. Sci.* **1**, 194 (1988).
61. De Vogt, P., Wegener, J.W.M., Brinkman, U.A.T. and Govers, H., *Sci. Total Environ.* **109/110**, 69 (1991).
62. Camps, F., Colomina, O., Messeguer, A. and Sanchez, F.J., *J. Liq. Chromatogr.* **9**, 23 (1986).
63. Taylor, R.B., Ochepke, N.A. and Wangboonskul, J., *J. Liq. Chromatogr.* **12**, 1645 (1989).
64. Klimes, J., Klimesova, V. and Waisser, K., *J. Chromatogr.* **595**, 334 (1993).
65. Valko, K. and Slegel, P., *J. Chromatogr.* **631**, 49 (1993).
66. Tsantili-Kakoulidou, A., Filippatos, E., Todooulou, O. and Papadaki-Valiraki, A., *J. Chromatogr. A* **654**, 43 (1993).
67. Hamoir, T. and Massart, D.L., *J. Chromatogr. A.* **673**, 1 (1994).
68. Reichelova, V., Albertioni, F and Liliemark, J., *J. Chromatogr. A* **667**, 37 (1994).
69. Gocan, S., Irimie, F. and Cimpan, G., *J. Chromatogr. A* **675**, 282 (1994).
70. Gago, F., Pastor, M., Pererez-Burtaguerio, J., Lopez, R., Alvarez-Builla, J. and Elguero, J., *Quant. Struct.-Act. Relat.* **13**, 165 (1994).
71. Knox, J.H. and Kaliszan, R., *J. Chromatogr.* **349**, 211 (1985).
72. Opperhuizen, A., Sinnige, T.L., Van der Steen, J.M.D. and Hutzinger, O., *J. Chromatogr.* **388**, 51 (1987).
73. Boyce, C.B.C. and Millborrow, B.V., *Nature (London)*, **208**, 537 (1965).
74. Soczewinski, E. and Wachtmeister, C.A., *J. Chromatogr.* **7**, 311 (1962).
75. Schoenmakers, P.J., Billiet, H.A.H., Tijssen, R. and De Galan, L., *J. Chromatogr.* **149**, 519 (1978).
76. Snyder, L.R., Dolan, J.W. and Gant, J.R., J. Chromatogr. **165**, 3 (1979).
77. Miyake, K., Mizuno, N. and Terada, H., *J. Chromatogr.* **439**, 227 (1988).
78. Sherblom, P.M. and Eganhouse, R.P., *J. Chromatogr.* **454**, 37 (1988).
79. Clark, C.R., Barksdale, J.M., Mayfield, C.A., Ravis, W.A. and DeRuiter J., *J. Chromatogr. Sci.* **28**, 83 (1990).
80. Taylor, R.B., Ochekpe, N.A. and Wangboonskul, J., *J. Liquid Chromatogr.* **12**, 1645 (1989).
81. Kaibara, A., Hohda, C., Hirata, N., Hirose, M. and Nakagawa, T. *Chromatographia* **29**, 275 (1990).
82. Johnson, B.P., Khaledi, M.G. and Dorsey, J.G., *Anal. Chem.* **58**, 2354 (1986).
83. Minick, D.J., Frenz, J.H., Patrick, M.A. and Brent, D.A., *J. Med. Chem.* **31**, 1923 (1988).
84. Michels, J.J. and Dorsey, J.D., J. Chromatogr. **499**, 435 (1990).
85. Hsieh, M.M. and Dorsey, J.G., *J. Chromatogr.* **631**, 63 (1993).
86. Park, J.H., Dallas, A.J., Chan, P. and Carr, P.W., *J. Chromatogr. A* **677**, 1 (1994).
87. Braumann, T., Genieser, H.-G., Lullmann, C. and Jastorff, B., *Chromatographia* **24**, 777 (1987).
88. Klaas, E.B., Horvath, C., Melander, W.R. and Nahum, A., *J. Chromatogr.* **203**, 65 (1981).
89. Schoenmakers, P.J., Billiet, H.A.H. and De Galan, L., *J. Chromatogr.* **218**, 261 (1981).
90. Janos, E., Cserhati, T. and Tyihak, E., *J. High Resol. Chromatogr. Chromatogr. Commun.* **5**, 634 (1982).
91. Cserhati, T., *Chromatographia* **18**, 318 (1984).
92. Pietrogrande, M.C., Bighi, C., Borea, P.A., Barbaro, A.M., Guerra, M.C. and Biagi, G.L., *J. Liq. Chromatogr.* **8**, 1711 (1985).
93. Wilson, I.D., *J. Chromatogr.* **354**, 99 (1986).
94. Cserhati, T., Bordas, B. and Osapoy, G., *Chromatographia* **23**, 184 (1987).
95. Kovacs-Hadady, K. and Szilagyi, J., *J. Chromatogr.* **553**, 459 (1991).
96. Dross, K.P., Mannhold, R. and Rekker, R.F., *Quant. Struct.-Act. Relat.* **11**, 36 (1992).

97. Dross, K., Sonntag, C. and Mannhold, R., *J. Chromatogr.* **639**, 287 (1993).
98. Biagi, G.L., Guerra, M.C., Barbaro, A.M., Barbieri, S., Recanatini, M. and Borea, P.A., *J. Liq. Chromatogr.* **13**, 913 (1990).
99. Biagi, G.L., Barbaro, A.M., Guerra, M.C., Borea, P.A. and Recanatini, M., *J. Chromatogr.* **504**, 163 (1990).
100. Biagi, G.L., Barbaro, A.M., Sapone, A. and Recanatini, M., *J. Chromatogr A* **662**, 341 (1994).
101. Biagi, G.L., Barbaro, A.M., Sapone, A. and Recanatini, M., *J. Chromatogr. A* **669**, 246 (1994).
102. Biagi, G.L., Barbaro, A.M. and Recanatini, M., *J. Chromatogr A* **678**, 127 (1994).
103. Kuchar, M., Kraus, E. and Jelinkova, M., *J. Chromatogr.* **557**, 399 (1991).
104. Cserhati, T., *J. Liq. Chromatogr.* **16**, 1805 (1993).
105. Tomlinson, E., Poppe, H. and Kraak, J.C., *Int. J. Pharm.* **7**, 225 (1981).
106. Kaliszan, R., *J. Chromatogr. Sci.* **22**, 362 (1984).
107. Horvath, Cs., Melander, W. and Molnar, I., *J. Chromatogr.* **125**, 126 (1976).
108. Schoenmakers, P.J., Billiet, H.A.H. and De Galan, L., *J. Chromatogr.* **185**, 179 (1979).
109. Antle, P.E., Goldberg, A.P. and Snyder, L.R., *J. Chromatogr.* **321**, 1 (1985).
110. Eisenbeis, F., *Ber. Bunsen-Ges. Phys. Chem.* **93**, 1019 (1989).
111. Nawrocki, J., *Chromatographia* **31**, 177 (1991).
112. Freiser, H.H., Nowlan, M.P. and Gooding, D.L., *J. Liquid Chromatogr.* **12**, 827 (1989).
113. Stadalius, M.A., Berus, J.S. and Snyder, L.R., *LC-GC* **6**, 494 (1988).
114. Minick, D.J., Brent, D.A. and Frenz, J., *J. Chromatogr.* **461**, 177 (1989).
115. Eadsforth, C.V., *Pest. Sci.* **17**, 311 (1986).
116. Leo, A., *J. Pharm. Sci.* **76**, 166 (1987).
117. Kaliszan, R., *Quant. Struct.-Act. Relat.* **9**, 83 (1990).
118. De Biasi, V., Lough, W.J. and Evans, M.B., *J. Chromatogr.* **353**, 279 (1986).
119. Bitteur, S. and Rosset, R., *J. Chromatogr.* **394**, 279 (1987).
120. Miyake, K., Kitaura, F., Mizuno, N. and Terada, H., *Chem. Pharm. Bull.* **35**, 377 (1987).
121. Bechalany, A., Rothlisberger, T., El Tayar, N. and Testa, B., *J. Chromatogr.* **473**, 115 (1989).
122. Tanaka, N., Ebata, T., Hashizume, K., Hosoya, K., Araki, M., Araki, T. and Kimata, K., *Abstracts of Papers, 13th Symposium on Column Liquid Chromatography*, Stockholm 1989; No. W-L-012.
123. Dawkins, J.V., Gaggot, N., Lloyd, L.L., McConville, J.A. and Warner, F.P., *J. Chromatogr.* **452**, 145 (1988).
124. Bechalany, A., Tsantili-Kakoulidou, A., El Tayar, N. and Testa, B., *J. Chromatogr.* **541**, 221 (1991).
125. Lambert, W.J. and Wright, L.A., *J. Chromatogr.* **464**, 400 (1989).
126. Yamaguchi, J. and Hanai, T., *Chromatographia* **27**, 371 (1989).
127. Sentell, K.B., Bornes, K.W. and Dorsey, J.G., *J. Chromatogr.* **455**, 95 (1988).
128. Ascah, T.L. and Feibush, B., *J. Chromatogr.* **506**, 357 (1990).
129. Othsu, Y., Shiojima, Y., Okumura, T., Koyama, J., Nakamura, K. and Nakata, O., *J. Chromatogr.* **481**, 147 (1989).
130. Hetem, M.J.J., De Haan, J.W., Claessens, H.A., Cramers, C.A., Deege, A. and Schomburg, G., *J. Chromatogr.* **540**, 53 (1991).
131. Schomburg, G., Deege, A., Bien-Vogelsang, U. and Kohler, J., *J. Chromatogr.* **287**, 27 (1983).
132. Engelhardt, H., Low, H., Eberhardt, W. and Mauss, M., *Chromatographia* **26**, 535 (1989).
133. Kaliszan, R., Blain, R.W. and Hartwick, R.A., *Chromatographia* **25**, 5 (1988).
134. Bien-Vogelsang, U., Deege, A., Figge, H., Kohler, J. and Schomburg, G., *Chromatographia* **19**, 170 (1984).
135. Haky, J.E., Vemulapalli, S. and Wieserman, L.F., *J. Chromatogr.* **505**, 307 (1990).
136. Knox, J.H. and Kaur, B., in: "*High Performance Liquid Chromatography*", Brown, P.R. and Hartwick, R.A. (Eds.), Wiley, New York, 1989; Chapter 4.
137. Weber, T.P. and Carr, P.W., *Anal. Chem.* **62**, 2620 (1990).
138. Rigney, M.P., Weber, T.P. and Carr, P.W., *J. Chromatogr.* **484**, 273 (1989).
139. Miyake, K., Kitaura, F., Mizuno, N. and Terada, H, *J. Chromatogr.* **389**, 47 (1987).
140. Seydel, J.K., Albores Velasco, M., Coats, E.A., Cordes, H.P., Kunz, B. nad Wiese, M., *Quant. Struct.-Act. Relat.* **11**, 205 (1992).
141. Thurnhofer, H., Schnabel, J., Betz, M., Lipka, G., Pidgeon, C. and Hauser, H., *Biochim. Biophys. Acta* **1064**, 275 (1991).
142. Pidgeon, C., Marcus, C. and Alvarez, F., in: "*Applications of Enzyme Biotechnology*", Baldwin T.O. and Kelly J.W. (Eds.), Plenum Press, New York, 1992; p. 201.
143. Kaliszan, R., Kaliszan, A. and Wainer, I.W., *J. Pharm. Biomed. Anal.* **11**, 505 (1993).
144. Kaliszan, R., Nasal, A. and Buciński, A., *Eur. J. Med. Chem.* **29**, 163 (1994).
145. Nasal, A., Sznitowska, M., Buciński, A. and Kaliszan, R., *J. Chromatogr.* **692**, 83 (1995).
146. Craig, P.N., in: "*Comprehensive Medicinal Chemistry*", Vol. 6, Hansch, C., Sammes, P.G. and Taylor, J.B. (Eds.), Pergamon Press, Oxford, 1990.

147. Ong, S., Liu, H., Qiu, X., Bhat, G. and Pidgeon, C., *Anal. Chem.* **67**, 755 (1995).
148. Seydel, J.K., Coats, E.G., Cordes, H.P. and Wiese, M., *Arch. Pharm.* **327**, 60 (1994).
149. Ong, S. and Pidgeon, C., *Anal. Chem.* **67**, 2119 (1995).
150. Kaliszan, R., Nasal, A. and Turowski, M., *Biomed. Chromatogr.* **9**, 211 (1995).
151. Kaliszan, R., Nasal, A. and Turowski, M., *J. Chromatogr. A*, **722**, 25 (1996).
152. Hansch, C. and Leo, A., "*Substituent Constants for Correlation Analysis in Chemistry and Biology*", Wiley, New York, 1979; Appendix II.
153. Kaliszan, R., Petrusewicz, J., Blain, R.W. and Hartwick, R.A., *J. Chromatogr.* **458**, 395 (1988).
154. Minick, D.J., Sabatka, J.J. and Brent, D.A., *J. Liquid Chromatogr.* **10**, 2565 (1987).
155. Gami-Yilinkou, R. and Kaliszan, R., *Chromatographia* **30**, 277 (1990).
156. Gami-Yilinkou, R. and Kaliszan, R., *Pol. J. Pharmacol. Pharm.* **44**, 515 (1992).
157. Gami-Yilinkou, R. and Kaliszan, R., *J. Chromatogr.* **550**, 573 (1991).
158. Terada, H., Kosuge, Y., Murayama, W., Nakaya, N., Nunogaki, Y. and Nunogaki, K.-I., *J. Chromatogr.* **400**, 343 (1987).
159. Berthod, A. and Armstrong, D.W., *J. Liq. Chromatogr.* **11**, 547 (1988).
160. Berthod, A. and Armstrong, D.W., *J. Liq. Chromatogr.* **11**, 567 (1988).
161. Altomare, C., Tsai, R.-S., El Tayar, N., Testa, B., Carotti, A., Cellamare, S. and De Benedetti, P.G., *J. Pharm. Pharmacol.* **43**, 191 (1991).
162. El Tayar, N., Marston, A., Bechalany, A., Hostettmann, K. and Testa, B., *J. Chromatogr.* **469**, 91 (1989).
163. Vallat, P., El Tayar, N., Testa, B., Slacanin, J., Marston, A. and Hostettmann, K., *J. Chromatogr.* **504**, 411 (1990).
164. Tsai, R.-S., El Tayar, N., Testa, B. and Ito, Y., *J. Chromatogr.* **538**, 119 (1991).
165. Ito, Y. and Bowman, R.L., *J. Chromatogr.* **147**, 221 (1978).
166. Ito, Y., *J. Chromatogr.* **188**, 33 (1980).
167. Ito, Y., *J. Chromatogr.* **207**, 161 (1981).
168. Ito, Y. and Oka, H., *J. Chromatogr.* **457**, 393 (1988).
169. Ito, Y. and Bowman, R.L., *Anal. Biochem.* **85**, 614 (1978).
170. Ito, Y., *J. Chromatogr.* **192**, 75 (1980).
171. Murayama, W., Kobayashi, T., Kosuge, Y., Yano, H., Nunogaki, Y. and Nunogaki, K., *J. Chromatogr.* **239**, 649 (1982).
172. Armstrong, D.W., Bertrand, G.L. and Berthod, A., *Anal. Chem.* **69**, 2513 (1988).
173. Gluck, S.J. and Martin, E.J., *Chromatography, Materials of the Chromatography International Symposium*, Tokyo, 1989, p. 291.
174. El Tayar, N., Tsai, R.-S., Vallat, P., Altomare, C. and Testa, B., *J. Chromatogr.* **556**, 181 (1991).
175. Berthod, A. and Armstrong, D.W., *J. Liq. Chromatogr.* **11**, 1187 (1988).
176. Gago, F., Alvarez-Builla, J. and Elguero, J., *J. Chromatogr.* **360**, 247 (1986).
177. Gago, F., Alvarez-Builla, J. Elguero, J. and Diez-Masa, J., *Anal. Chem.* **59**, 921 (1987).
178. Berthod, A., *Analysis* **18**, 352 (1990).
179. Berthod, A., Menges, R.A. and Armstrong, D.W., *J. Liq. Chromatogr.* **19**, 2769 (1992).

12 Combination of Chromatography and Chemometrics in Modeling of Pharmacological Activity of Solutes

12.1. Introduction

Fundamental processes of drug action and the processes that are the basis of chromatographic separations have much in common. Biological processes of absorption, distribution, excretion and receptor activation are dynamic in nature as are the solute's distribution processes in chromatography. None of the essential pharmacological or chromatographic processes (excepting metabolism) implies breaking of existing bonds in a drug (solute) molecule or the formation of new bonds. The same basic intermolecular interactions determine the behaviour of chemical compounds in both biological and chromatographic environments [1]. However, the extreme stereospecificity of biological interactions has been the feature differentiating them from interactions determining regular chromatographic separations.

Modern techniques and procedures of high-performance liquid chromatography (HPLC) and capillary electrophoresis (CE) allow for inclusion of biomolecules as active components of chromatographic systems. This approach is now termed "biochromatography". The extreme complexity of biological systems limits rational design of an individual chromatographic system that would directly mimic a given biological system. On the other hand, chromatography is a unique method that can readily yield a great amount of diversified, precise and reproducible data. It can be presumed that chemometric processing of appropriately designed and selected sets of chromatographic data can reveal systematic information regarding the xenobiotics studied [2].

The ability of chromatography to readily produce a great amount of precise and reproducible data is an advantage that QSRR has over other types of quantitative structure–property relationship studies. In a chromatographic process all of the experimental conditions can be kept constant and the solute structure becomes the single independent variable in the physicochemical system.

There are numerous reports regarding biochromatographic separations. Also, QSRR analysis of various chromatographic data has been the subject of many publications [1–4]. However, the combination of biochromatography and chemometrics is a research strategy very recently introduced and developed [2,5,6]. Such a combined approach is demonstrated here based on recent publications by the authors and co-workers to provide information of direct relevance to molecular pharmacology and drug design.

12.2. Hydrophobicity Parameter Determined on the Immobilized Artificial Membrane (IAM) Stationary Phase in Correlation Studies with Bioactivity

The hydrophobic effect is assumed to be one of the driving forces for passive diffusion of xenobiotics through biological membranes and drug-receptor binding. For this reason, much effort has been devoted to develop reliable and convenient hydrophobicity measuring systems. All such systems yield quantities reflecting the degree of "phobia" of the solutes against aqueous phase and "philia" towards nonpolar species (lipophilicity). However, a complex property like hydrophobicity depends on solute environment and, thus, individual measuring systems produce data that highlight specifically different aspects of what is called (more or less intuitively) hydrophobicity [7].

The 1-octanol-water partition system is the common reference system for determination of relative hydrophobicity [8]. Expressed as the logarithm of the partition coefficient, log *P*, this measure is the most widely employed in medicinal chemistry and molecular pharmacology. Large compilations of log *P* data have been developed using data determined experimentally by the standard "shake-flask" method and from theoretical calculations based on the fragmental method, CLOGP data [8–10].

Because of the tediousness and poor reproducibility of hydrophobicity parameters determined by the slow equilibrium methods, researchers turned their attention towards reversed-phase high-performance liquid chromatography (RP HPLC) as a tool for the convenient assessment of relative hydrophobicities within a series of solutes. The RP HPLC systems commonly used in general hydrophobicity determinations employ octanol-like or hydrocarbonaceous stationary phase materials [11].

The RP HPLC approach has been relatively successful in duplicating log *P* data derived by traditional techniques. However, if the reason to use the chromatographic system is to model processes in the biophase, then the components of the chromatographic and the biological systems must be comparable. Thus, an RP HPLC system which is used to model transport through biological membranes should be composed of an aqueous phase and an organized phospholipid layer.

Miyake *et al.* [12] derived HPLC hydrophobicity parameters employing a column of silica gel coated physically with dipalmitoyl phosphatidylcholine (DPPC). Leaving aside inconveniences regarding their preparation and stability, the systems with DPPC adsorbed onto silica most probably do not emulate the lipid dynamics of biological membranes, because the adsorbed lipids are not organized in a manner similar to natural (or artificial) membranes.

Recently, a new RP HPLC stationary phase material, the immobilized artificial membrane stationary phase (IAM), became available [13]. This phase more closely models natural membranes which are composed of lipids with a polar headgroup and two nonpolar chains. In the IAM phase, a lipid molecule, lecithin, is covalently bound to propylamine silica forming confluent monolayers of immobilized membrane lipids. In one form of the IAM, unreacted propylamine moieties are end-capped with methylglycolate. Only one of the alkyl chains is linked to the propylamine silica surface, and the immobilized lipid headgroups protrude away from the stationary phase surface. These charged moieties are the first contact site between solutes and IAM. Relationships between $\log P$ and hydrophobicity parameters determined on the IAM columns have been discussed in Chapter 11.

Preliminary studies reported encouraging correlations between HPLC retention parameters determined on an IAM and human skin permeation data for a short series of alcohols and steroids [13,14]. The same biological data showed poor correlation with HPLC retention parameters determined on a regular hydrocarbonaceous reversed-phase HPLC columns.

Retention parameters determined on an immobilized artificial membrane column, $\log k'_{\mathrm{IAM}}$, were found [15] to correlate better with several pharmacokinetic parameters of β-adrenolytics than the logarithms of apparent 1-octanol-buffer partition coefficient determined by Hinderling *et al.* [16] by the "shake-flask" method, $\log k'_{\mathrm{SF4}}$, and the logarithms of HPLC capacity factor determined in an 1-octanol-like chromatographic system, $\log k'_{\mathrm{C}_{18}}$. In Table 12.1 the statistical parameters of respective equations relating pharmacokinetic parameters to $\log k'_{\mathrm{IAM}}$, $\log k'_{\mathrm{SF}}$ and $\log k'_{\mathrm{C18}}$ are given. In every case but one, the statistics of the regressions describing bioactivity in terms of $\log k'_{\mathrm{IAM}}$ appears better than that provided by $\log k'_{\mathrm{SF}}$ and by another HPLC hydrophobicity measure, $\log k'_{\mathrm{C}_{18}}$.

Table 12.1 Statistical parameters of regression equations relating pharmacokinetic parameters to experimental hydrophobicity parameters: n – number of data points; R – correlation coefficient; s – standard error of estimate. (Reprinted with permission from Kaliszan, R., Nasal, A. and Buciński, A., *Eur. J. Med. Chem.* 29, 163. Copyright Publications Elsevier, Paris.)

Pharmacokinetic parameter	*Hydrophobicity descriptor*								
	$\log k'_{IAM}$ [a]			$\log k'_{SF}$ [b]			$\log k'_{C18}$ [c]		
	n	R	s	n	R	s	n	R	s
$\log t_T$ [d]	9	0.83	0.44	11	0.84	0.46	9	0.74	0.55
$\log f_b$ [e]	13	0.83	0.37	14	0.80	0.40	10	0.76	0.47
$\log r_A$ [f]	6	0.95	0.15	6	0.90	0.21	6	0.88	0.23
$\log K_{BC}$ [g]	10	0.90	0.18	11	0.86	0.22	9	0.76	0.26
$\log V_{uss}$ [h]	9	0.86	0.23	11	0.84	0.30	9	0.73	0.31
$\log K_P$ [i]	10	0.87	0.45	13	0.90	0.43	9	0.85	0.46
$\log r$ [j]	9	0.82	0.58	11	0.93	0.38	9	0.95	0.32

[a] Logarithm of HPLC capacity factor determined on an immobilized artificial membrane column at pH 7.0; [b] logarithm of apparent n-octanol/buffer pH 7.4 partition coefficient determined by the shake-flask method; [c] logarithm of HPLC capacity factor determined in an octanol-like chromatographic system at pH 7.4; [d] logarithm of the fraction of drug bound and unbound to tissue; [e] logarithm of fraction of drug bound in plasma; [f] logarithm of ratio of fraction of drug bound and unbound to albumin; [g] logarithm of true red cell partition coefficient; [h] logarithm of steady state volume of distribution referenced to the unbound drug in plasma; [i] logarithm of partition coefficient of drug between plasma protein and plasma water; [j] logarithm of ratio of the fraction of drug nonrenally and renally eliminated; [a] data from Kaliszan *et al.* [15]; [b–j] data from Hinderling *et al.* [16].

In the same study [15] logarithms of the reciprocals of phenothiazine antihemolytic concentrations, log $1/C$, taken from Seeman [17] were related to: (1) the standard 1-octanol-water partition coefficient obtained from Craig [10] which were corrected for ionization, $\log D$; and, (2) to $\log k'_{\mathrm{IAM}}$ (Table 12.2). The respective regression equations were:

$$\log 1/C = 1.126(\pm 0.174)\log k'_{\mathrm{IAM}} + 3.266(\pm 0.266); \qquad (12.1)$$
$$n = 8; \quad r = 0.935; \quad s = 0.102;$$

$$\log 1/C = 0.405(\pm 0.130)\log D + 3.662(\pm 0.416); \qquad (12.2)$$
$$n = 7; \quad r = 0.812; \quad s = 0.170.$$

The $\log k'_{\mathrm{IAM}}$ data for a series of steroid hormones [18] are given in Table 12.3 along with the reference 1-octanol-water partition parameter, $\log P$, taken from El Tayar *et al.* [19]. The human skin/water partition coefficient, $\log P_m$, and human skin permeation coefficient, $\log K_P$, were taken from Scheuplein *et al.* [20].

The relationship between $\log K_P$ and $\log k'_{\mathrm{IAM}}$ for the steroids is given by the equation

$$\log K_P = 1.77(\pm 0.22)\log k'_{\mathrm{IAM}} - 10.19(\pm 0.37); \qquad (12.3)$$
$$n = 10; \quad r = 0.942; \quad p < 10^{-5},$$

where p is the significance level of the regression equation. The extent of correlation provided by Eq. (12.3) is illustrated in Fig. 12.1 (upper part).

Table 12.2 pK_a values, hydrophobicity parameters and antihaemolytic activity of a series of phenothiazine neuroleptics.

Drug	pK_a^{a}	*log* P^{b}	*log* D^{c}	*CLOGP*[d]	*CLOGD*[e]	*log* k'_{IAM}[f]	*log* $1/C^{g}$
Chlorpromazine	9.30	5.35	3.05	5.20	2.90	1.440	4.72
Fluphenazine	8.10	4.36	3.23	5.90	4.77	1.401	5.00
Perphenazine	7.80	4.20	3.34	5.57	4.71	1.373	4.82
Prochloperazine	8.10			6.15	5.02	1.782	5.22
Promazine	9.40	4.55	2.15	4.28	1.88	1.171	4.55
Promethazine	9.10			4.65	2.55	1.195	
Propiomazine	9.10			5.00	2.90	1.265	
Thioridazine	9.50	5.90	3.40	6.42	3.92	1.747	5.30
Trifluoperazine	8.10	5.03	3.90	6.48	5.35	1.754	5.22
Trifluopromazine	9.20	5.19	2.99	5.53	3.33	1.474	4.96

[a] Acidity constant; [b] logarithm of experimental n-octanol/water partition coefficient for neutral forms determined by the shake-flask method; [c] logarithm of experimental n-octanol/water partition coefficient corrected for ionization at pH 7.0; [d] logarithm of theoretical n-octanol/water partition coefficient for neutral form of solute calculated by fragmental method; [e] logarithm of theoretical n-octanol/water partition coefficient corrected for ionization at pH 7.0; [f] logarithm of HPLC capacity factor determined on an immobilized artificial membrane column; [g] logarithm of reciprocal of haemolytic concentration; [a,b,d] data taken from Craig [10]; [c,e,f] data calculated or determined by Kaliszan, *et al.* [15]; [g] data taken from Seeman [17].

Table 12.3 Human skin permeability coefficients (log K_P, log P_m) and hydrophobicity parameters (log k'_{IAM}, log P) of a group of steroid hormones. (Reprinted from Nasal, A., Sznitowska, M., Buciński, A. and Kaliszan, R., *J. Chromatogr. A* 692, 83, 1995 with kind permission of Elsevier Science-NL, Sara Burgerhartstraat 25, 1055 KV Amsterdam, The Netherlands.)

Steroid hormone	*log* K_P [a]	*log* P_m [a]	*log* k'_{IAM} [b]	*log* P [c]
Aldosterone	−9.08	0.83	0.649	1.08
Cortisone	−8.56	0.93	0.772	1.42
Deoxycorticosterone	−6.90	1.57	1.730	2.88
Estradiol	−7.08	1.66	2.179	2.69
Estriol	−7.95	1.36	1.368	2.47
Estrone	−6.00	1.66	2.001	2.76
Hydrocortisone	−9.08	0.85	0.843	1.53
Pregnenolone	−6.38	1.70	2.124	3.13
Progesterone	−6.38	2.01	2.199	3.70
Testosterone	−6.95	1.36	1.693	3.31

[a] K_P (cm/s) and P_m according to Scheuplein *et al.* [20]; [b] data determined on and immobilized artificial membrane column by Nasal *et al.* [18]; [c] logarithms of *n*-octanol/water partition coefficient according to El Tayar *et al.* [19].

The corresponding equation which describes $\log K_P$ in terms of $\log P$ has the form

$$\log\ K_P = 1.18(\pm 0.20)\log P - 10.39(\pm 0.53); \qquad (12.4)$$

$$n = 10; \quad r = 0.900; \quad p < 10^{-4}.$$

Correlation between the observed $\log K_P$ data and the data calculated by Eq. (12.4) is illustrated in Figure 12.1 (lower part).

Good statistical quality of both equation (12.3) and (12.4) can be explained by the fact that the two hydrophobicity parameters, $\log k'_{IAM}$ and $\log P$, are evidently intercorrelated ($r = 0.911$). However, as evident from Fig. 12.1 the $\log k'_{IAM}$ is a better predictor of skin permeation than $\log P$.

For the group of hormones analyzed a highly significant correlation between the skin/water partition coefficient, $\log P_m$, and hydrophobicity parameter determined on the IAM column, $\log k'_{IAM}$, has also been observed:

$$\log P_m = 0.64(\pm 0.06)\ \log k'_{IAM} + 0.40(\pm 0.10); \qquad (12.5)$$

$$n = 10; \quad r = 0.966; \quad p < 10^{-5}.$$

The corresponding equation describing $\log P_m$ in terms of $\log P$ has the form:

$$\log P_m = 0.42(\pm 0.06)\ \log P + 0.33(\pm 0.17); \qquad (12.6)$$

$$n = 10; \quad r = 0.917; \quad p < 10^{-4}.$$

Previously Raykar *et al.* [21] reported correlation between $\log P$ and $\log P_m$. The high statistical quality of Eqs. (12.5) and (12.6) proves that both $\log k'_{IAM}$ and $\log P$ can be used to evaluate the partitioning of steroids between water and striatum corneum. Again, the $\log k'_{IAM}$ seems to be more reliable for this purpose than $\log P$.

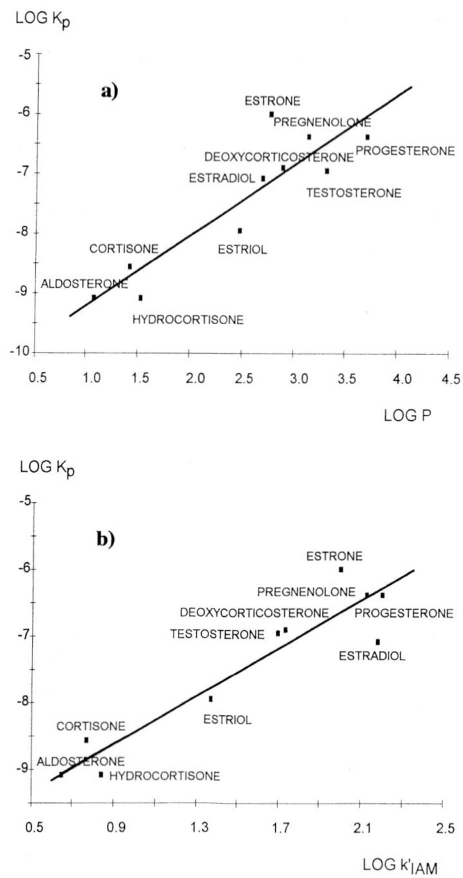

Figure 12.1. Relationships between logarithm of human skin permeation coefficient of steroids according to Scheuplein *et al.* [20] and logarithm of capacity factor determined on an immobilized artificial membrane column (upper plot) and logarithm of 1-octanol-water partition coefficient (lower plot) (Reprinted from Nasal, A., Sznitowska, M., Buciński, A. and Kaliszan, R., *J. Chromatogr. A* 692, 83, 1995 with kind permission of Elsevier Science-NL, Sara Burgerhartstraat 25, 1055 KV Amsterdam, The Netherlands.)

12.3. Prediction and Description of Drug-Human Serum Albumin Binding from the Chemometric Analysis of Biochromatographic Retention Parameters

Chemometric analysis of enantiospecific biochromatographic data obtained on protein-based stationary phases can be used to relate solute structure and properties (topography) of binding site(s) on the immobilized protein. Biochromatography on an immobilized protein readily produces a substantial amount of precise and reliable binding-related data which is necessary for chemometric analysis. An example of this approach is the biochromatographic measurement of the enantiospecific interactions of benzodiazepine (BDZ) derivatives with HSA [22].

A series of BDZ derivatives (Fig. 12.2) was subjected to HPLC analysis [6,23]. The set of 22 compounds included nine achiral solutes, nine racemic mixtures and four single enantiomers. The compounds were chromatographed on an HSA chiral stationary phase (CSP). The mobile phase was based on sodium dihydrogen phosphate/disodium hydrogen phosphate pH 6.90, modified with 5% (v/v) propanol.

In the analysis of quantitative structure-enantiospecific retention relationships (QSERR) for chiral solutes, the logarithms of capacity factors corresponding to both first peak, $\log k'_P$, and the second peak, $\log k'_M$, were considered; retention parameters

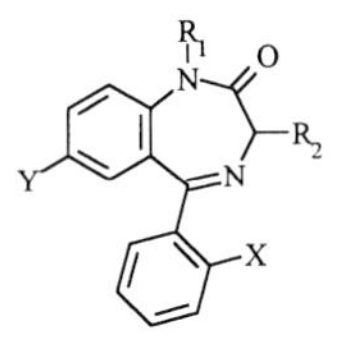

Compounds 1-14

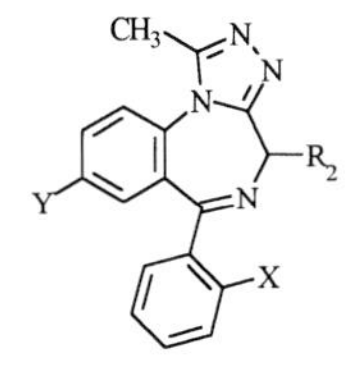

Compounds 15-18

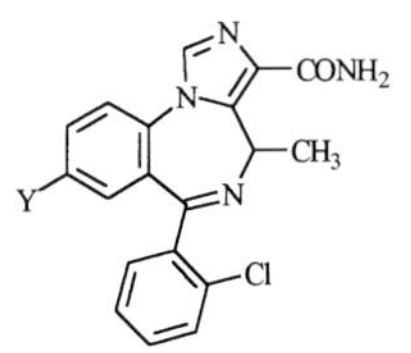

Compounds 19 and 20

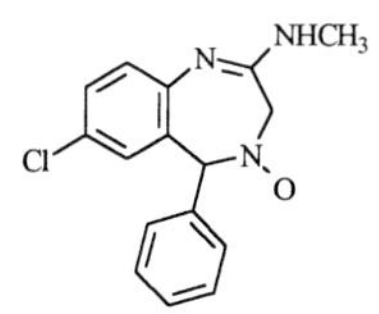

Compound 21

	Compound	R_1	R_2	X	Y
1	Desmethyldiazepam	H	H	H	Cl
2	Delorazepam	H	H	Cl	Cl
3	Nitrazepam	H	H	H	NO_2
4	Clonazepam	H	H	Cl	NO_2
5	Oxazepam	H	OH	H	Cl
6	Oxazepam Hemisuccinate	H	$OCO(CH_2)_2COOH$	H	Cl
7	Lorazepam	H	OH	Cl	Cl
8	(S)-23 0983/001	H	CH_3	Cl	F
9	(R,S)-11 3128/002	H	CH_3	Cl	NO_2
10	Diazepam	CH_3	H	H	Cl
11	Flunitrazepam	CH_3	H	F	NO_2
12	(S)-14 8935/000	CH_3	CH_3	Cl	NH_2
13	Temazepam	CH_3	OH	H	Cl
14	Lormetazepam	CH_3	OH	Cl	Cl
15	(R,S)-Alprazolam, 4-OH	OH	-	H	Cl
16	Alprazolam	H	-	H	Cl
17	Triazolam	H	-	Cl	Cl
18	(S)-11 5073/000	H	-	F	Cl
19	(S)-23 1117/000	-	-	-	F
20	(R,S)-23 3880/000	-	-	-	Cl
21	Chlordiazepoxide	*See above*			

Figure 12.2. Benzodiazepine test solutes chromatographed on an immobilized human serum albumin column.

of the achiral agents, $\log k'_{AC}$, were analyzed independently. The structure of the compounds was characterized by means of the following parameters (Fig. 12.3) [6]: C_3 – quantum chemically calculated electron excess charge on carbon C_3; P_{SM}– the excess charge difference between the hydrogen atom at C_3 and the most negatively charged atom in the other substituent at C_3 multiplied by the distance (in Å) between these two atoms; W– width of the molecule along the phenyl substituent; f_{X+Y} – the sum of the hydrophobic constants of the substituent at position 7 in the fused benzene ring plus that of the substituent at position 2′ of the phenyl substituent. Numerical data are collected in Table 12.3.

At first, a highly significant regression equation was obtained relating retention of the second eluting enantiomer, which is assumed to bind in the M conformation, $\log k'_M$, and retention of the first eluting enantiomer, which is assumed to bind in the P conformation, $\log k'_P$ (Fig. 12.4):

$$\log k'_M = 1.097(\pm 0.162) \log k'_P + 0.547(\pm 0.067)\, P_{SM} - 0.149; \qquad (12.7)$$
$$n = 8; \quad R = 0.980; \quad F = 61; \quad p < 3 \times 10^{-4};$$

where F is the value of the F-test of significance.

The retention of the first eluted enantiomer was described as:

$$\log k'_P = 0.183(\pm 0.079)\, f_{X+Y} - 0.278(\pm 0.055)\, W + 2.479; \qquad (12.8)$$
$$n = 13; \quad R = 0.845; \quad F = 12.5; \quad p < 2 \times 10^{-3}.$$

This equation can be interpreted such that retention (or binding) takes place at site which contains a hydrophobic pocket and some steric restrictions.

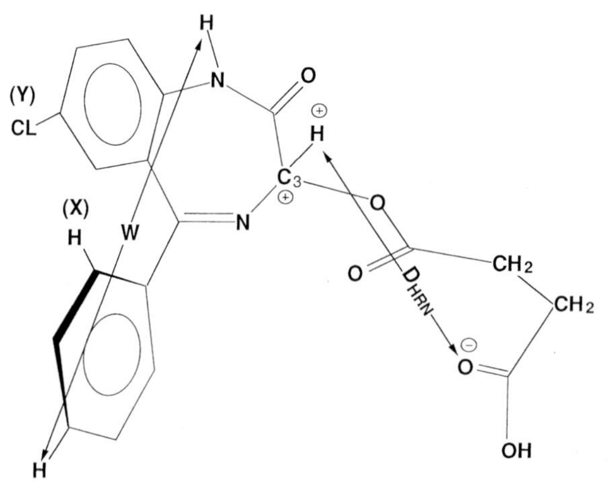

Figure 12.3. Structural descriptors of benzodiazepines used in quantitative structure-enantiospecific retention relationship (QSERR) equations. See text for explanation.

Table 12.4 Logarithms of chromatographic capacity factors determined on a human serum albumin HPLC column and structural parameters from molecular modeling of a series of benzodiazepine derivatives (5). (Reprinted with permission from Kaliszan, R., Noctor, T.A.G. and Wainer, I.W., *Mol. Pharmacol.* 42, 512, 1992. Copyright American Society for Pharmacology and Experimental Therapeutics.)

No[a]	$log\ k'_P$[b]	$log\ k'_M$[c]	$log\ k'_{AC}$[d]	P_{SM}[e]	f_{X+Y}[f]	C_3[g]	W[h]
1			0.8645	0.0849	1.05	0.1035	9.30
2	0.8512	1.8938		1.8635	1.05	0.2785	8.74
3			0.6243	0.0703	0.06	0.0960	8.54
4			0.4857	0.0609	0.20	0.0882	9.63
5			0.7679	0.0680	0.77	0.0966	8.67
6			1.0614	0.0635	1.76	0.0977	8.69
7			1.0969	0.0634	1.05	0.0979	8.59
8			1.1216	0.0578	1.05	0.0933	9.56
9	0.7672	0.9745		0.6120	1.76	0.2388	9.76
10	0.8068	0.9360		0.5953	1.76	0.2425	8.71
11	0.6561	1.0261		0.7049	1.05	0.2451	8.60
12	0.5224	1.1793		0.6113	1.05	0.2353	9.49
13	0.3892			0.0675	−0.18	0.0549	10.0
14	0.6628			0.0600	1.19	0.0624	8.64
15	0.7193	0.7193		0.0633	0.77	0.0651	8.70
16	0.2648	0.4533		0.5862	1.05	0.3169	10.5
17			0.4200	0.0784	1.05	0.1722	10.3
18			0.6243	0.0730	1.76	0.1725	10.3
19	0.3838			0.0484	1.19	0.1379	10.2
20	0.7404			0.0320	1.19	0.1309	9.20
21	1.0523	1.1156		0.0441	1.76	0.1184	9.22
22	0.9715	1.3992		1.0745	1.05	0.0916	8.14

[a] Compounds are numbered as in Fig. 12.2; [b] retention parameter of the first eluting enantiomer; [c] retention parameter of the second eluting enantiomer; [d] retention parameter of achiral solutes; [e] submolecular polarity parameter (see Fig. 12.3); [f] sum of hydrophobic constants of substituents X and Y (see Fig. 12.3); [g] electron excess charge of carbon C_3 of the diazepine system; [h] molecular width in Å.

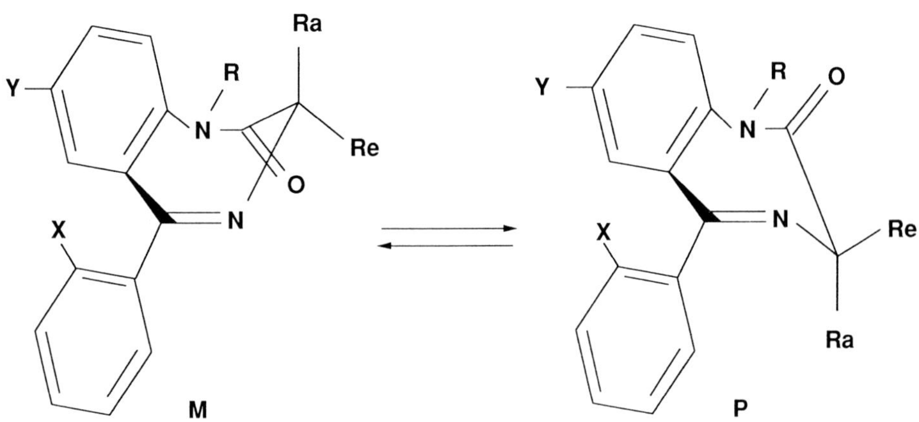

Figure 12.4. Conformations *M* and *P* of benzodiazepines.

The retention of the second eluting enantiomer of benzodiazepines is described by the equation

$$\log k'_M = 0.835(\pm 0.154)\, P_{\mathrm{SM}} + 0.364(\pm 0.199)\, f_{X+Y} - 2.690(\pm 0.932)\,\mathrm{C}_3 + 0.556;$$
$$n = 8; \quad R = 0.938; \quad F = 10; \quad p < 0.03 \qquad (12.9)$$

This result indicates that binding occurs at a site which contains both a hydrophobic pocket and a positively charged area. The cationic region of the binding site produces an attractive interaction with the submolecular dipole quantified by P_{SM} and a repulsive interaction with the excess positive charge at C_3.

The retention of achiral benzodiazepines was described as

$$\log k'_{\mathrm{AC}} = 0.374(\pm 0.096)\, f_{\mathrm{X+Y}} - 7.068(\pm 1.663)\,\mathrm{C}_3 + 1.221; \qquad (12.10)$$
$$n = 9; \quad R = 0.889; \quad F = 11; \quad p < 0.01.$$

Since for achiral benzodiazepines the parameter P_{SM} is zero, Eq. (12.9) indicates that these compounds predominantly bind in the M conformation as reported in the literature [24].

Based on Eqs. (12.7)–(12.10), a model was proposed [5] for the structural requirements of the two postulated modes of benzodiazepine binding to HSA (Fig. 12.5). According to Eq. (12.8), benzodiazepines appear to bind within hydrophobic cavities, and substituents at N_1, C_2 and C_5 would then provide spatial orientation of

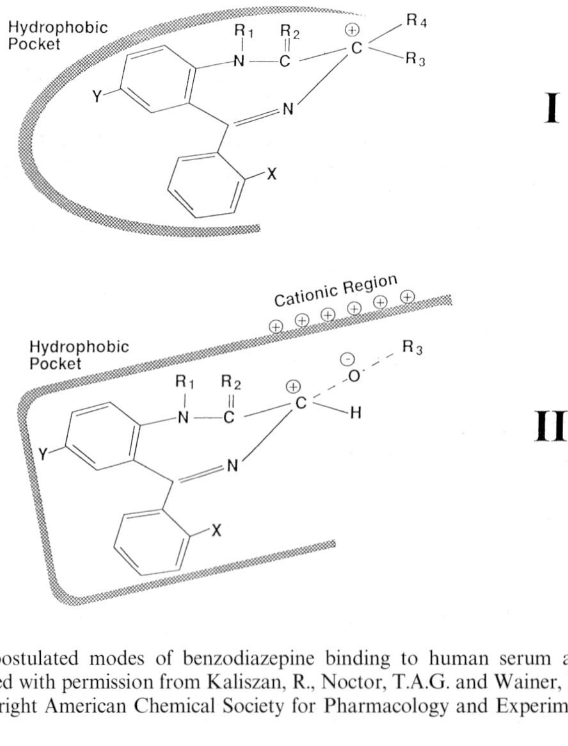

Figure 12.5. Two postulated modes of benzodiazepine binding to human serum albumin. See text for discussion. (Reprinted with permission from Kaliszan, R., Noctor, T.A.G. and Wainer, I.W., *Mol. Pharmacol.* 42, 512, 1992. Copyright American Chemical Society for Pharmacology and Experimental Therapeutics.)

the BDZ molecules within this cavity. Steric limitations suggest that the hydrophobic cavity has definite boundaries. In addition, the steric features at the stereogenic center, carbon C_3, appear to play no role in this binding mode (Fig. 12.5I).

The binding mode of the second eluting enantiomer, as described by Eq. (12.9), involves hydrophobic and electrostatic interactions. Thus, in addition to a hydrophobic cavity, there must be a cationic region in close proximity (Fig. 12.5II). For the benzodiazepines in the *M* conformation, the electrostatic repulsion between the excess positive charge on carbon C_3 and the cationic site on the protein surface appears to be more than offset by the attraction of a negatively charged atom within the substituent at C_3 and the same area. In the case of the lesser retained enantiomer, the electrostatic repulsion between carbon C_3 and the cationic area and the steric hindrance due to the *P* conformation may prohibit binding at this site.

12.4. Binding Site for Basic Drugs on α_1-Acid Glycoprotein in view of QSRR Analysis of Biochromatographic Data

α_1-Acid glycoprotein (AGP) is a serum protein which mainly binds basic drugs [25]. The prevalent view is that AGP has only one common drug binding which binds drugs through hydrophobic and electrostatic interactions [26,27]. However, neither $\log P$ [26,28] nor pK_a [29] can account for binding differences within small sets of tested drugs.

Capacity factors, $\log k'_{\mathrm{AGP}}$, were determined for 52 basic drugs of diverse chemical structure and pharmacological activity on an AGP-CSP [30–32]. The column was packed with human AGP chemically bound to 5 μm silica particles; the mobile phase was isopropanol-phosphate buffer (0.1 M, pH 6.5) 5:95% (v/v); eluent flow rate was 0.5 ml/min and UV detection was at $\lambda = 215$ nm. The capacity factors, k'_{AGP}, were determined using the retention of sodium nitrate peak as the marker of column dead volume. The $\log k'_{\mathrm{IAM}}$ parameters for the set of agents studied were also determined using an IAM column.

Molecular modeling was employed to determine the solute structural parameters which were important to describe their interactions with AGP. The following parameters were used: electron excess charge (N_{ch}) and the surface area (S_T) of a triangle having one vertex on the aliphatic nitrogen and the two remaining vertices on the extremely positioned atoms in the drug molecule (Fig. 12.6) [31]. Structural parameters from molecular modeling and retention data determined on AGP and IAM columns are given in Table 12.5.

The QSRR equation relating retention on chemically immobilized AGP to a hydrophobicity measure, $\log k'_{\mathrm{IAM}}$, electron excess charge on aliphatic nitrogen, N_{ch}, and a size parameter, S_T, of drugs has the form

$$\begin{aligned}\log k'_{\mathrm{AGP}} &= 0.6577(\pm 0.0402)\log k'_{\mathrm{IAM}} + 3.342(\pm 0.841)\, N_{\mathrm{ch}} \\ &\quad - 0.0081(\pm 0.0030)\, S_T + 1.688(\pm 0.245); \qquad (12.11)\\ &n = 49;\quad R = 0.929;\quad s = 0.163;\quad F = 92;\quad p < 10^{-5}.\end{aligned}$$

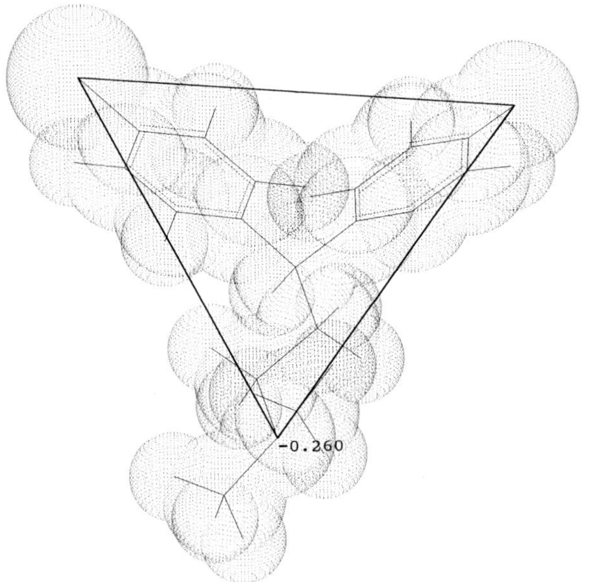

Figure 12.6. Structural descriptors of basic drugs binding to human α_1-acid glycoprotein (AGP) used in quantitative structure-retention relationships (QSRR) studies. See text for explanation.

Table 12.5 Logarithms of chromatographic capacity factors determined on and α_1-acid glycoprotein HPLC column, log k'_{AGP}, and on an immobilized artificial membrane column, log k'_{IAM}, and structural parameters of a series of basic drugs [31]. (Reprinted with permission from Kaliszan, R., Nasal, A. and Turowski, M., *Biomed. Chromatogr.* 9, 211, 1995. Copyright John Wiley & Sons Limited.)

Drug name	*log* k'_{AGP}	*log* k'_{IAM}	*CLOGP*[a]	N_{ch}[b]	S_T[c]
Antagonists of H_1 *histamine receptor*					
Antazoline	1.154	1.043	4.25	−0.285	25.23
Chlorpheniramine	1.202	1.055	2.73	−0.260	30.92
Chloropyramine	1.431	1.330	3.56	−0.268	31.11
Cinnarizine	2.148	2.250	6.14	−0.248	32.18
Dimethindene	1.382	1.194	3.42	−0.267	29.76
Diphenhydramine	1.140	1.006	3.36	−0.266	32.78
Isothipendyl	1.580	1.210	3.93	−0.260	20.41
Ketotifen	1.459	1.168	3.56	−0.245	25.54
Mepyramine	1.113	0.935		−0.272	32.81
Pheniramine	0.926	0.602	2.02	−0.260	27.37
Pizotifen	1.898	1.588		−0.247	29.13
Promethazine	1.833	1.508	4.65	−0.257	20.73
Tripelennamine	1.066	0.887	2.85	−0.264	30.37
Triprolidine	1.185	1.084	3.47	−0.240	24.02
Tymazoline	1.306	1.204		−0.287	17.07

Table 12.5 (*Continued*).

Drug name	$log\ k'_{AGP}$	$log\ k'_{IAM}$	$CLOGP^a$	$N_{ch}{}^b$	S_T^c
Antagonists of H_2 histamine receptor					
Cimetidine	0.482	−0.271	0.21	−0.195	18.11
Famotidine	0.731	−0.271	−0.57	−0.161	27.44
Metiamide	0.517	−0.301	0.38	−0.193	19.19
Nizatidine	0.460	−0.368		−0.275	32.01
Ranitidine	0.600	−0.016	0.27	−0.262	31.86
Roxatidine	0.773	0.359	2.66	−0.279	28.99
Antagonists of beta-adrenoceptors					
Acebutolol	0.676	0.602	1.61	−0.300	49.35
Alprenolol	1.490	0.918	2.59	−0.299	25.99
Atenolol	0.499	−0.146	−0.11	−0.298	20.36
Betaxolol	0.838	0.994	2.17	−0.300	27.00
Bisoprolol	0.694	0.646	1.69	−0.297	45.07
Bopindolol	1.940	0.456	4.86	−0.301	38.39
Bupranolol	0.981	0.269		−0.291	21.76
Carteolol	0.706	−0.146	1.17	−0.289	24.88
Celiprolol	0.700	0.723	1.66	−0.304	37.28
Cicloprolol	0.735	1.012		−0.298	42.28
Dilevalol	1.106	1.272	2.18	−0.295	27.49
Esmolol	0.649	0.646	1.53	−0.299	30.30
Metoprolol	0.564	0.434	1.20	−0.301	22.66
Nadolol	0.606	0.269		−0.292	29.90
Nifenalol	0.639	0.269	1.16	−0.300	12.19
Oxprenolol	1.210	0.586	1.62	−0.299	28.60
Pindolol	0.870	0.586	1.65	−0.301	27.79
Practolol	0.509	−0.067	0.78	−0.301	22.74
Propranolol	1.612	1.340	2.75	−0.300	28.77
Sotalol	0.516	−0.146	0.23	−0.178	19.40
Timolol	0.696	0.385	1.63	−0.290	15.60
Agonists and antagonists of alpha-adrenoceptors					
Cirazoline	1.082	0.940	3.27	−0.288	16.78
Clonidine	0.847	0.410		−0.278	12.32
Doxazosin	1.798	1.983	3.77	−0.307	30.64
Moxonidine	0.528	−0.067		−0.286	13.35
Naphazoline	1.092	0.895	3.83	−0.295	17.61
Phentolamine	1.264	1.340	3.68	−0.289	24.30
Prazosin	1.390	1.594	2.16	−0.308	31.47
Tiamenidine	0.808	0.434		−0.284	9.96
Tramazoline	1.315	1.123	2.49	−0.285	21.22
UK-14, 304	0.831	0.269		−0.276	15.74

[a]Logarithm of *n*-octanol/water partition coefficient calculated by the fragmental method taken from Craig [10]; [b]electron excess charge on aliphatic nitrogen; [c]area (in Å) of the size/shape triangle (see Fig. 12.6).

Of 52 drugs studied, three (triprolidine, cimetidine and bopindolol) did not fit Eq. (12.11). The reason may be the uncertainty in identifying the atom equivalent to the aliphatic nitrogen with regard to electrostatic interactions.

There are two aspects of QSRR equations: their predictive potency and their physical meaning. Predictive power of Eq. (12.11) is illustrated in Fig. 12.7. Certainly, for some agents the predicted strength of chromatographically measured interactions with AGP differs from the observed one. However, there is no better known means to evaluate the actual binding more properly.

Equation (12.11) may be useful in drug design as a first approximation of relative binding of an agent to AGP without a need to perform biological experiments. It can also indicate relative differences in AGP binding among individual drugs and help to predict probability of pharmacokinetic interactions of two simultaneously administered drugs.

Equation (12.11) comprises a term ($\log k'_{\mathrm{IAM}}$) which cannot be obtained from computational chemistry and thus, to use it one needs some amounts (however minute) of the existing substance. There is a hydrophobicity parameter which can be calculated from structural formula, *i.e.*, CLOGP [8–10]. For 38 drugs for which both $\log k'_{\mathrm{IAM}}$ and CLOGP were available the correlation between them was $r = 0.850$. Replacing $\log k'_{\mathrm{IAM}}$ in Eq. (12.11) by CLOGP decreases correlation but the resulting regression equation remains significant and thus informative. Most probably, the

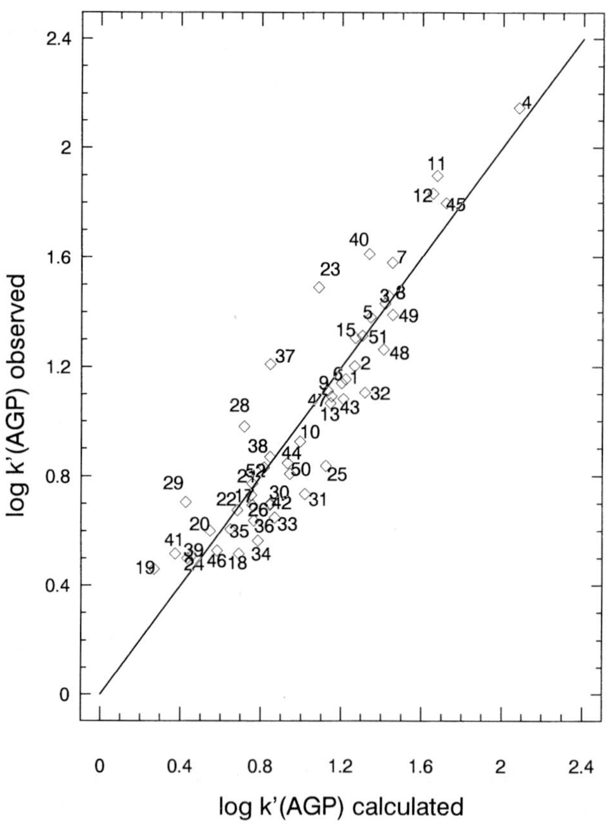

Figure 12.7. Logarithm of capacity factor determined on an α_1-acid glycoprotein (AGP) column against the value calculated by Eq. (12.11). (Reprinted with permission from Kaliszan, R., Nasal, A. and Turowski, M., *Biomed. Chromatogr.* 9, 211, 1995. Copyright John Wiley & Sons Limited.)

predictive power of CLOGP could increase if a correction due to ionization at actual pH was introduced. The problem in individual cases is, however, to acquire the required pK_a values.

The QSRR equation derived here and the reported [25–29] qualitative characteristics of the mode of binding of xenobiotics by AGP allow one for an indirect identification of structural features of the binding site of basic drugs (Fig. 12.8). The site can be modeled by a conical pocket. Its wall (internal surface) contains lipophilic regions at the base of the cone. There is an anionic region close to the spike of the cone. Protonated aliphatic nitrogen guides drug molecules towards the anionic region. Hydrophobic hydrocarbon fragments of the interacting drugs provide anchoring in the lipophilic region(s) of the binding site. There is a steric restriction

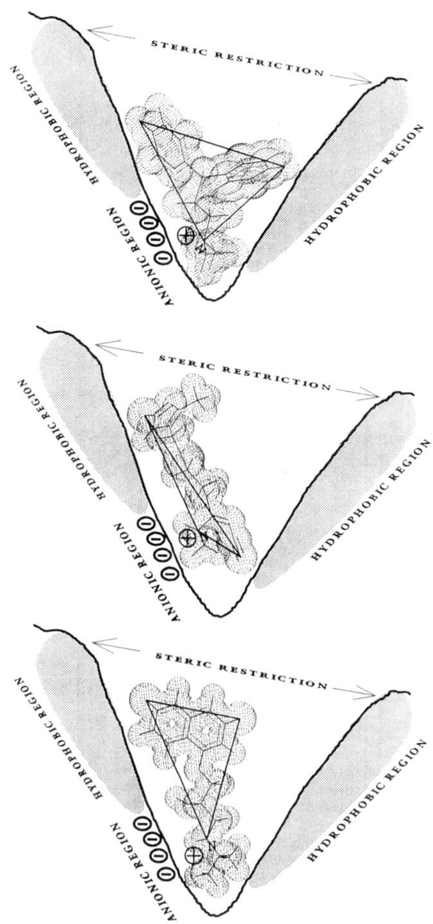

Figure 12.8. Mode of binding of basic drugs by α_1-acid glycoprotein. See text for discussion. (Reprinted with permission from Kaliszan, R., Nasal, A. and Turowski, M., *Biomed. Chromatogr.* 9, 211, 1995. Copyright John Wiley & Sons Limited.)

for the molecule to plunge into the binding site. Asymmetric charge distribution accounts for the observed enantioselectivity of binding to AGP. Similarly, asymmetric distribution of charge (positive) in one of the two benzodiazepine binding sites on HSA accounted for the formerly discussed stereoselectivity of binding.

12.5. Comparison of Chromatographic and Pharmacological Properties of Drug Solutes

It is commonly assumed that pharmacological (toxicological) effects of xenobiotics are determined by the processes of transport (usually passive diffusion) of an agent to the site of action and by its intermolecular interactions with receptors. Both kinds of biological phenomena depend on compounds hydrophobicity. Of course, nobody may question the requirements of a specific complementarity of the interacting entities. However, Ehrlich's paraphrase of an old alchemical rule "corpora non agunt nisi soluta" into "corpora non agunt nisi fixata" calls in mind the hydrophobic binding as that providing an intimate contact without breaking of the existing chemical bonds nor the formation of new bonds.

It could be presumed that the only indispensable and satisfactory condition to employ a given hydrophobicity measure in QSAR studies is its good predictive potency. However, it looks sometimes like for most researchers the main reason to use a given chromatographic parameter of hydrophobicity is that it conforms to the 1-octanol-water $\log P$ scale.

On the other hand, there are undoubtedly differences in the properties of specific parts of the body penetrated by a given drug and in the binding properties of pharmacological and "silent" receptors. Thus, it can be argued that for prediction of the net effects of complex pharmacokinetic and pharmacodynamic processes, information extracted from diversified partition chromatographic data may be more useful than information expressed by an individual, one-dimensional hydrophobicity scales.

Recently, instead of a more or less arbitrary choice of a single hydrophobicity scale the strategy is recommended to make use of information dispersed over diversified, yet often highly intercorrelated, hydrophobicity data. To extract that systematic information modern multivariate chemometric methods of data analysis are employed.

If there is no single, unique, universal, continuous, unequivocally defined and pharmacologically distinguished hydrophobicity scale then there is no reason to prefer the information on properties of solutes provided by an individual RP HPLC system over the information gained from measurements performed in another chromatographic system. On the contrary, systematic information extracted from diversified retention data may be more appropriate for prediction of the net effects of complex pharmacokinetic and pharmacodynamic processes than information based on an individual one-dimensional hydrophobicity scale. To extract the systematic information from diversified (yet often highly intercorrelated) sets of data, the modern multivariate chemometric methods of data analysis must be employed. All the reproducible retention data provide information on structure of the solutes. Also the data discarded in traditional methods of reduction of dimensionality of the data sets comprise information of relevance for structural characterization of individual compounds of the series studied. By discarding retention data outside the range of linearity of the dependence of $\log k'$ on the volume fraction of organic modifier, information on specific properties of the solutes is lost. Valid structural information is also lost if reliable retention parameters which do not correlate with the 1-octanol-water $\log P$ data are excluded from considerations.

Musumara and co-workers [33–37] demonstrated the application of principal component analysis (PCA) of thin-layer chromatographic (TLC) data for predicting of toxicity of xenobiotics. Wold and co-workers [38,39] reported multivariate parametrization of amino acid properties based on TLC data obtained in diversified separation systems. The TLC data considered were R_f values determined in seven straight-phase isocratic systems differing with respect to the type of stationary phase and composition of the eluent. The PCA of the data matrix comprising the R_f parameters, along with Van der Waals volume and molecular mass, resulted in two significant principal components. The principal components were shown to posses predictive capacity and explained about 70% of the variance in the literature data on pharmacological activity of a series of oligopeptides. The differences in pharmacological activities considered by Wold and co-workers [38,39] were quantitative, *i.e.*, all the agents elicited the same effect but of varying magnitude. In the case of the imidazol(in)e drugs the aim was to qualitatively differentiate bioactive substances by means of diversified chromatographic data. Small peptides have also been the subject of chromatographic and factorial analysis reported by Cserhati [40].

An application of chromatographic data for classification of individual members of a chemically closely related but pharmacologically diverse group of solutes demonstrates informative potential of such data [41]. The test group of solutes were imidazol(in)e circulatory drugs exerting their activity through the so called adrenergic receptors of the alpha type. Structural formulae of the drugs are given in Fig. 11.3.

There are no distinctive structural features which would justify separation of the agents into classes. Yet, pharmacologically are the drugs ascribed to two main classes: those binding preferentially to α_1 adrenoceptor and others possessing higher affinity to the α_2 adrenoceptor [42]. As a consequence of these differences opposite circulatory effects are observed.

Large sets of polycratic RP HPLC capacity factors determined at pH 7.0 on an ODS column and at pH 10.9 on a PBCA column (see Chapter 11) were considered. The polycratic parameters were extrapolated to $\log k'_w$ values. None of the two chromatographic hydrophobicity scales allowed for a pharmacologically consistent ordering of the solutes. Moreover, there were no regular distribution patterns of the drugs on the plane determined by $\log k'_w$ generated in the two systems studied (Fig. 12.9).

The principal component analysis (PCA) was performed on a matrix of capacity factors for 18 drug solutes determined in 21 RP HPLC systems. The first principal component (PC1) accounted for 60.5% and the second principal component (PC2) accounted for 18.9% of the variance in capacity factors considered.

The inputs provided to the principal components PC1 and PC2 by individual drugs ("object scores") were calculated. In Fig. 12.10 are displayed the positions of the drugs on the plane spanned by two principal component axes. The objects in

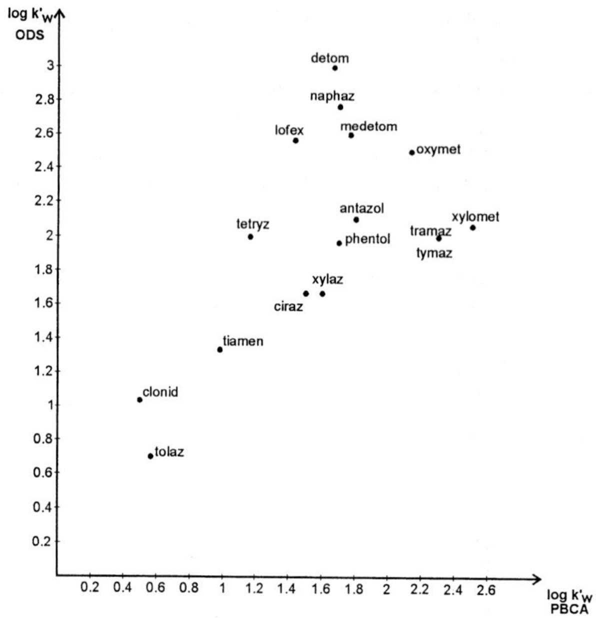

Figure 12.9. Distribution of imidazol(in)e drugs on the plane determined by the logarithms of capacity factors extrapolated to pure buffer as the eluent. The respective retention data were obtained using the ODS (pH 7.0) and the PBCA (pH 10.9) columns. (Reprinted from Gami-Yilinkou, R. and Kaliszan, R., *J. Chromatogr.* 550, 573, 1991 with kind permission of Elsevier Science-NL, Sara Burgerhartstraat 25, 1055 KV Amsterdam, The Netherlands.)

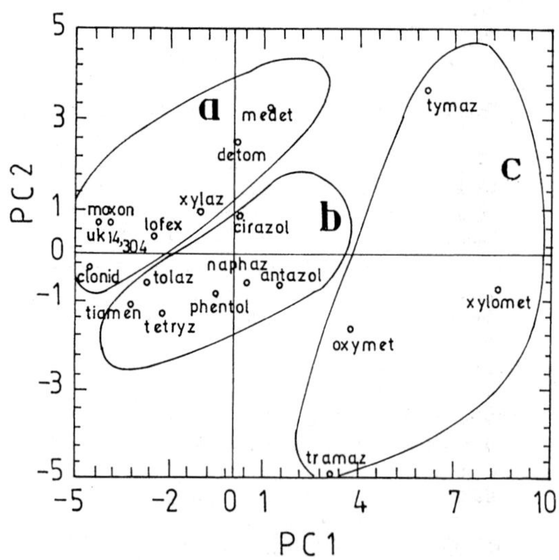

Figure 12.10. Two-dimensional scatter plot of the scores by individual imidazol(in)es in the two first principal components, PC1 and PC2, obtained from principal component analysis of chromatographic data. (Reprinted from Gami-Yilinkou, R. and Kaliszan, R., *J. Chromatogr.* 550, 573, 1991 with kind permission of Elsevier Science-NL, Sara Burgerhartstraat 25, 1055 KV Amsterdam, The Netherlands.)

Fig. 12.10 can be grouped into three clusters: **a**, **b** and **c**. Such a grouping due to the retention behaviour correlates well with the established pharmacological classification of the solutes. Pharmacology textbooks classify unequivocally the agents belonging to cluster **a** as selective agonists of α_2 adrenoceptor, whereas those belonging to cluster **c** are considered as pure α_1 agonists. It can be demonstrated that imidazolines belonging to cluster **b** possess affinity to both subtypes of alpha adrenoceptors.

For a set of 22 drugs (Fig. 12.11) known to modify physiological effects of histamine, the HPLC retention data were determined employing two reversed-phase columns, seven compositions of methanol–buffer eluent and three pH values of the buffer [43]. The $\log k'_w$ values were determined by extrapolation of the respective data in the five RP HPLC systems studied. Chemometric analysis by the PCA method of the 5 × 22 matrix of $\log k'_w$ data obtained for 22 drug solutes in 5 HPLC systems allowed for extraction of two main principal components which accounted

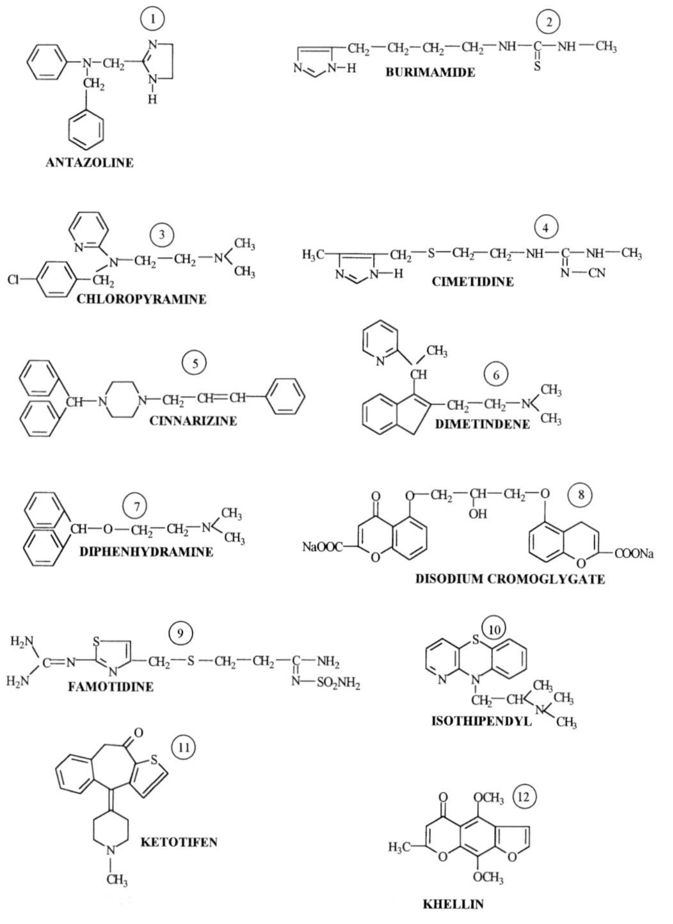

Figure 12.11.

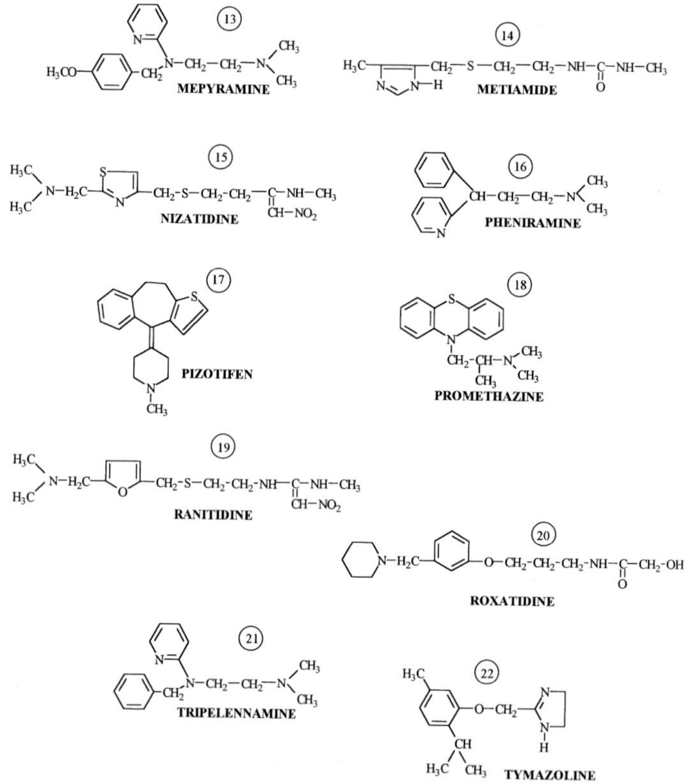

Figure 12.11. Chemical structure of antihistamine drugs.

for 96% of total data variance. The distribution of individual drugs on the plane determined by the two first principal component axes forms the patterns (Fig. 12.12) which are in good agreement with the established pharmacological classification of the structurally diverse compounds.

There are two clear-cut clusters of agents in Fig. 12.12 and three solutes show distinctive chromatographic behaviour. Undobtedly, compounds **2**, **4**, **9**, **14**, **15**, **19** and **20** form a compact cluster **a** of drugs unequivocally classified as antagonists of the H_2 type of histamine receptor. Among the compounds forming cluster **b** there are diverse chemicals classified and employed clinically as typical antagonists of H_1 receptor. Interestingly, to the cluster **b** belong also ketotifen (**11**) and pizotifen (**17**). Although clinical indications for ketotifen is prophylaxis of bronchial asthma and for pizotifen it is migraine, both are reported to possess antagonist activity towards H_1 receptors. Compounds **8** and **12** clearly do not belong to cluster **a** nor **b**. Khellin (**12**) is a myolytic agent and its derivative, disodium cromoglycate (**8**), decreases the liberation of histamine which accompanies the antigen-antibody reaction. None of the drug is known to interfere with the H_1 nor H_2 receptors. Cinnarizine (**5**) has diverse sites of actions, including some antihistaminic properties. Central effects

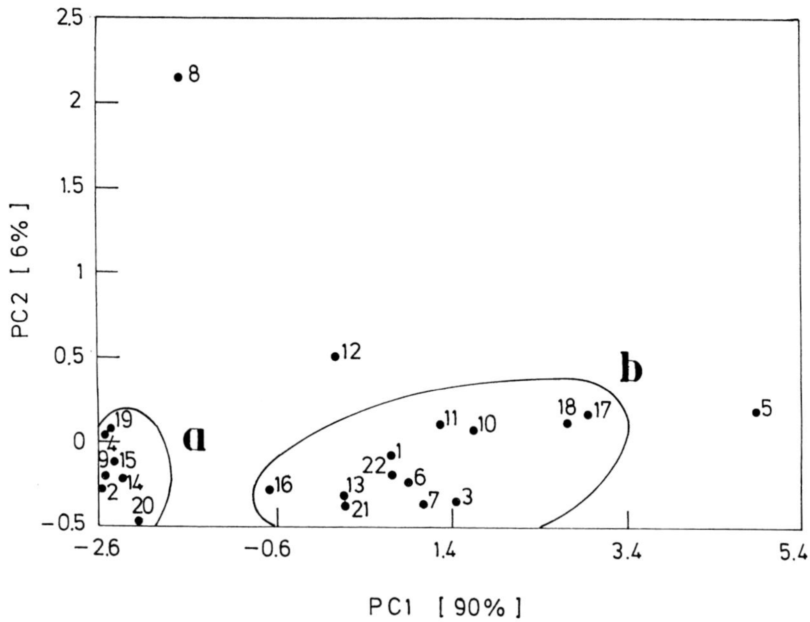

Figure 12.12. Inputs by individual drug solutes in two first principal components, PC1 and PC2, obtained from analysis of log k'_w data determined in five reversed-phase high-performance liquid chromatographic systems. The numbers denote the drugs: 1 – antazoline, 2 – burimamide, 3 – chlorpyramine, 4 – cimetidine, d – cinnarizine, 6 – dimethindene, 7 – diphenhydramine, 8 – disodium cromoglycate, 9 – famotidine, 10 – isothipendyl, 11 – ketotifen, 12 – khellin, 13 – mepyramine, 14 – metiamide, 15 – nizatidine, 16 – pheniramine, 17 – pizotifen, 18 – promethazine, 19 – ranitidine, 20 – roxatidine, 21 – tripelenamine and 22 – tymazoline. (Reprinted from Gami-Yilinkou, R., Nasal, A. and Kaliszan, R., *J. Chromatogr.* 633, 57, 1993 with kind permission of Elsevier Science-NL, Sara Burgerhartstraat 25, 1055 KV Amsterdam, The Netherlands.)

ascribed to the drug are certainly connected with its high hydrophobicity (and thus brain barrier permeation), as is reflected by a high score in the first principal component.

Systematic information extracted by the PCA from a set of retention data determined in various RP HPLC systems suffices for differentiation of imidazol(in)e and antihistamine drugs in accordance with their pharmacological classification. That result demonstrates usefulness of diverse chromatographic data for characterization of solutes and, thus, for prediction of their properties. By using multivariate methods of data analysis the informative capacity can be exploited of the retention data so far neglected in the retention-bioactivity relationship studies. These neglected data (which reflect generally an ability of a solute to take part in specific intermolecular interactions with stationary and/or mobile phase) can provide structural information which does not manifest itself within a set of capacity factors changing regularly with changes of chromatographic conditions. Providing that the data are reproducible, the more diverse is the set of capacity factors considered the more reliable information on solute properties can be extracted. From the view point of application of RP HPLC in studies of QSAR it appears more productive to collect a representative set of diverse retention parameters than to concentrate all the efforts on determination of a universal chromatographic measure of hydrophobicity.

12.6. Concluding remarks

Chemometric comparisons clearly demonstrate the advantages of biochromatographic measures of hydrophobicity determined by HPLC on an immobilized artificial membrane stationary phase. The parameter $\log k'_{IAM}$ can conveniently be obtained and appears to be a better predictor of some kinds of pharmacological activity. It does not mean that $\log P$ and other chromatographic hydrophobicity parameters are useless. Hydrophobicity is a multidimensional property and $\log k'_{IAM}$ simply increases our means to quantify different features of this biologically important phenomenon.

QSRR equations (Eqs. (12.7)–(12.11)) account in strict statistical terms for the properties qualitatively ascribed to the drug binding sites on HSA and AGP. Thus, pharmacologically relevant information can be assessed based on these indirectly derived models. Of course, it would be interesting to relate these working models to the structures of the drug-HSA and drug-AGP crystals once they are resolved. It must be stressed, however, that while crystallographic analysis helps to visualize supposed active sites on biomacromolecules, the pictorial presentation is of limited use in predicting the binding of agents of a given chemical structure. Reliable predictions of activity and the determination of the required structural properties for the design of new drugs are facilitated by the derivation of good quantitative structure pharmacological activity relationships. To derive such relationships one needs larger sets of numerically expressed biological (biorelevant) data. Such are the retention data determined by means of the HPLC systems containing biomolecules. A combination of biochromatography and chemometrics appears a promising strategy in biological research.

References

1. Kaliszan, R., "*Quantitative Structure-Chromatographic Retention Relationships*", Wiley, New York, 1987.
2. Kaliszan, R., *Chemom. Intell. Lab. Syst.* **24**, 89 (1994).
3. Kaliszan, R., *Anal. Chem.* **64**, 619A (1992).
4. Kaliszan, R., *Crit. Rev. Anal. Chem.* **16**, 323 (1986).
5. Kaliszan, R., Noctor, T.A.G. and Wainer, I.W., *Mol. Pharmacol.* **42**, 512 (1992).
6. Wainer, I.W., Kaliszan, R. and Noctor, T.A.G., *J. Pharm. Pharmacol.* **45** (Suppl. 1), 367 (1993).
7. Kaliszan, R., *Adv. Chromatogr.* **33**, 147 (1993).
8. Hansch, C. and Leo, A., "*Substituent Constants for Correlation Analysis in Chemistry and Biology*", Wiley, New York, 1979.
9. Rekker, R.F., "*The Hydrophobic Fragmental Constant*", Elsevier, Amsterdam, 1977.
10. Craig, P.N., in: "*Comprehensive Medicinal Chemistry*", Vol. 6, Hansch C., Sammes P.G. and Taylor J.B. (Eds.), Pergamon Press, Oxford, 1990.
11. Kaliszan, R., *J. Chromatogr. A* **656**, 417 (1993).
12. Miyake, K., Kitaura, F., Mizuno, N. and Terada, H., *J. Chromatogr.* **389**, 47 (1987).
13. Thurnhofer, H., Schnabel, J., Betz, M., Lipka, G., Pidgeon, C. and Hauser, H., *Biochim. Biophys. Acta* **1064**, 275 (1991).
14. Pidgeon, C., Marcus, C. and Alvarez, F., in: "*Applications of Enzyme Biotechnology*", Baldwin, T.O. and Kelly, J.W. (Eds.), Plenum Press, New York, 1992; p. 201.

15. Kaliszan, R., Nasal, A. and Buciński, A., *Eur. J. Med. Chem.* **29**, 163 (1994).
16. Hinderling, P.H., Schmidlin, O. and Seydel, J.K., *J. Pharmacokin. Biopharm.* **12**, 263 (1984).
17. Seeman, P., *Pharmacol. Rev.* **24**, 583 (1972).
18. Nasal, A., Sznitowska, M., Buciński, A. and Kaliszan, R., *J. Chromatogr.* **692**, 83 (1995).
19. El Tayar, N., Tsai, R.-S., Testa, B., Carrupt, P.-A., Hansch, C. and Leo, A., *J. Pharm. Sci.* **80**, 744 (1991).
20. Scheuplein, R.J., Blank, H., Brauner, G.J. and MacFarlane, D., *J. Invest. Dermatol.* **52**, 63 (1969).
21. Raykar, P.V., Fung, M.-C. and Anderson, B.D., *Pharm. Res.* **5**, 140 (1988).
22. Domenici, E., Bertucci, C., Salvadori, P., Felix, G., Cahagne, I., Motellier, S. and Wainer, I.W., *Chirality* **2**, 263 (1990).
23. Noctor, T.A.G., Phaam, C.D., Kaliszan, R. and Wainer, I.W., *Mol. Pharmacol.* **42**, 506 (1992).
24. Alebic-Kolbah, T., Kajfez, F., Rendic, S., Sunjic, V., Konowal, A. and Snatzke, G., *Biochem. Pharmacol.* **28**, 2457 (1979).
25. Kremer, J.M.H., Wilting, J. and Janssen, L.H.M., *Pharmacol Rev.* **40**, 1 (1988).
26. Schley, J., *J. Pharm. Pharmacol.* **39**, 132 (1987).
27. Soltes, L., Sebille, B. and Szalay, P., *J. Pharm. Biomed. Anal.* **12**, 1295 (1994).
28. Glaser, H. and Kriegelstein, J., *Naunyn-Schmiedebergs Arch. Pharmacol.* **265**, 321 (1970).
29. Goolkasian, D.L., Slaughter, R.L., Edwards, D.J. and Lalka, D., *Eur. J. Clin. Pharmacol.* **25**, 413 (1983).
30. Nasal, A., Radwańska, A., Ośmialowski, K., Buciński, A., Kaliszan, R., Barker, G.E., Sun, P. and Hartwick, R.A., *Biomed. Chromatogr.* **8**, 125 (1994).
31. Kaliszan, R., Nasal, A. and Turowski, M., *Biomed. Chromatogr.* **9**, 211 (1995).
32. Kaliszan, R., Nasal, A. and Turowski, M., *J. Chromatogr.* **722**, 25 (1996).
33. Musumarra, G., Scarlata, G., Romano, G. and Clementi, S., *J. Anal. Toxicol.* **7**, 286 (1983).
34. Musumarra, G., Scarlata, G., Romano, G., Clementi, S. and Wold, S., *J. Chromatogr. Sci.* **22**, 286 (1984).
35. Musumarra, G., Scarlata, G., Crima, G., Romano, G., Palazzo, S., Clementi, S. and Giulietti, G., *J. Chromatogr.* **295**, 31 (1984).
36. Musumarra, G., Scarlata, G., Crima, G., Romano, G., Palazzo, S., Clementi, S. and Giulietti, G., *J. Chromatogr.* **350**, 151 (1985).
37. Musumarra, G., Scarlata, G., Romano, G., Capello, G., Clementi, S. and Giulietti, G., *J. Anal. Toxicol.* **11**, 154 (1987).
38. Wold, S., Eriksson, L., Hellberg, S., Jonsson, J., Sjostrom, M., Skageberg, B. and Wikstrom, C., *Can. J. Chem.* **66**, 1814 (1987).
39. Eriksson, L., Jonsson, J., Hellberg, S., Sjostrom, M. and Wold, S., *Quant. Struct.-Act. Relat.* **7**, 144 (1988).
40. Cserhati, T., *J. Chromatogr.* **600**, 149 (1992).
41. Gami Yilinkou, R. and Kaliszan, R., *J. Chromatogr.* **550**, 573 (1991).
42. Bowman, W.C. and Rand, M.J., "*Textbook of Pharmacology*", Second Edition, Blackwell Scientific Publications, Oxford, 1980; pp. 11.31–11.32 and 23.47–23.49.
43. Gami-Yilinkou, R., Nasal, A. and Kaliszan, R., *J. Chromatogr.* **633**, 57 (1993).

Index